INTERNATIONAL
WILDLIFE
ENCYCLOPEDIA

THIRD EDITION

Volume 9

HAR–JAC

Marshall Cavendish Corporation
99 White Plains Road
Tarrytown, New York 10591–9001

Website: www.marshallcavendish.com

© 2002 Marshall Cavendish Corporation

Library of Congress Cataloging-in-Publication Data

Burton, Maurice, 1898-
    International wildlife encyclopedia / [Maurice Burton, Robert Burton] .-- 3rd ed.
        p. cm.
    Includes bibliographical references (p.).
    Contents: v. 1. Aardvark - barnacle goose -- v. 2. Barn owl - brow-antlered deer -- v. 3. Brown bear - cheetah -- v. 4. Chickaree - crabs -- v. 5. Crab spider - ducks and geese -- v. 6. Dugong - flounder -- v. 7. Flowerpecker - golden mole -- v. 8. Golden oriole - hartebeest -- v. 9. Harvesting ant - jackal -- v. 10. Jackdaw - lemur -- v. 11. Leopard - marten -- v. 12. Martial eagle - needlefish -- v. 13. Newt - paradise fish -- v. 14. Paradoxical frog - poorwill -- v. 15. Porbeagle - rice rat -- v. 16. Rifleman - sea slug -- v. 17. Sea snake - sole -- v. 18. Solenodon - swan -- v. 19. Sweetfish - tree snake -- v. 20. Tree squirrel - water spider -- v. 21. Water vole - zorille -- v. 22. Index volume.
    ISBN 0-7614-7266-5 (set) -- ISBN 0-7614-7267-3 (v. 1) -- ISBN 0-7614-7268-1 (v. 2) -- ISBN 0-7614-7269-X (v. 3) -- ISBN 0-7614-7270-3 (v. 4) -- ISBN 0-7614-7271-1 (v. 5) -- ISBN 0-7614-7272-X (v. 6) -- ISBN 0-7614-7273-8 (v. 7) -- ISBN 0-7614-7274-6 (v. 8) -- ISBN 0-7614-7275-4 (v. 9) -- ISBN 0-7614-7276-2 (v. 10) -- ISBN 0-7614-7277-0 (v. 11) -- ISBN 0-7614-7278-9 (v. 12) -- ISBN 0-7614-7279-7 (v. 13) -- ISBN 0-7614-7280-0 (v. 14) -- ISBN 0-7614-7281-9 (v. 15) -- ISBN 0-7614-7282-7 (v. 16) -- ISBN 0-7614-7283-5 (v. 17) -- ISBN 0-7614-7284-3 (v. 18) -- ISBN 0-7614-7285-1 (v. 19) -- ISBN 0-7614-7286-X (v. 20) -- ISBN 0-7614-7287-8 (v. 21) -- ISBN 0-7614-7288-6 (v. 22)
    1. Zoology -- Dictionaries.   I. Burton, Robert, 1941-   . II. Title.

    QL9 .B796 2002
    590'.3--dc21

                                                          2001017458

Printed in Malaysia
Bound in the United States of America

07 06 05 04 03 02 01   8 7 6 5 4 3 2 1

Brown Partworks
Project editor: Ben Hoare
Associate editors: Lesley Campbell-Wright, Rob Dimery, Robert Houston, Jane Lanigan, Sally McFall, Chris Marshall, Paul Thompson, Matthew D. S. Turner
Managing editor: Tim Cooke
Designer: Paul Griffin
Picture researchers: Brenda Clynch, Becky Cox
Illustrators: Ian Lycett, Catherine Ward
Indexer: Kay Ollerenshaw

Marshall Cavendish Corporation
Editorial director: Paul Bernabeo

Authors and Consultants

Dr. Roger Avery, BSc, PhD (University of Bristol)

Rob Cave, BA (University of Plymouth)

Fergus Collins, BA (University of Liverpool)

Dr. Julia J. Day, BSc (University of Bristol), PhD (University of London)

Tom Day, BA, MA (University of Cambridge), MSc (University of Southampton)

Bridget Giles, BA (University of London)

Leon Gray, BSc (University of London)

Tim Harris, BSc (University of Reading)

Richard Hoey, BSc, MPhil (University of Manchester), MSc (University of London)

Dr. Terry J. Holt, BSc, PhD (University of Liverpool)

Dr. Robert D. Houston, BA, MA (University of Oxford), PhD (University of Bristol)

Steve Hurley, BSc (University of London), MRes (University of York)

Tom Jackson, BSc (University of Bristol)

E. Vicky Jenkins, BSc (University of Edinburgh), MSc (University of Aberdeen)

Dr. Jamie McDonald, BSc (University of York), PhD (University of Birmingham)

Dr. Robbie A. McDonald, BSc (University of St. Andrews), PhD (University of Bristol)

Dr. James W. R. Martin, BSc (University of Leeds), PhD (University of Bristol)

Dr. Tabetha Newman, BSc, PhD (University of Bristol)

Dr. J. Pimenta, BSc (University of London), PhD (University of Bristol)

Dr. Kieren Pitts, BSc, MSc (University of Exeter), PhD (University of Bristol)

Dr. Stephen J. Rossiter, BSc (University of Sussex), PhD (University of Bristol)

Dr. Sugoto Roy, PhD (University of Bristol)

Dr. Adrian Seymour, BSc, PhD (University of Bristol)

Dr. Salma H. A. Shalla, BSc, MSc, PhD (Suez Canal University, Egypt)

Dr. S. Stefanni, PhD (University of Bristol)

Steve Swaby, BA (University of Exeter)

Matthew D. S. Turner, BA (University of Loughborough), FZSL (Fellow of the Zoological Society of London)

Alastair Ward, BSc (University of Glasgow), MRes (University of York)

Dr. Michael J. Weedon, BSc, MSc, PhD (University of Bristol)

Alwyne Wheeler, former Head of the Fish Section, Natural History Museum, London

Contents

HARVESTING ANT

Harvesting ants exhibit a strict division of labor within a colony. Small workers gather and transport seeds to the nest. Here the seeds are husked and crushed by larger workers, later to be fed to developing larvae.

HARVESTING ANTS ARE named for their habit of collecting seeds and grains and storing them in large quantities in their underground nests. These ants are found in the drier regions of the subtropics and some of them live in deserts. In southern Europe, Africa and Asia they are represented by *Messor*, *Pheidole* and other genera, while in subtropical North America ants of the genera *Pogonomyrmex* and *Ischnomyrmex* have adopted a similar mode of life.

Harvesting ants are polymorphic social insects, that is, they are divided into specialized forms or castes, each of which carries out a particular function within the colony. The ordinary workers are quite small and normally proportioned, but there are also larger workers and certain individuals, the largest ants of all, in which the jaws are enlarged and the heads are proportionally enormous to accommodate huge jaw muscles. Each caste has its own role in the colony.

Villagelike colonies

The nests of colonies of harvesting ants are large, sometimes forming a mound 20 or 30 feet (6–9 m) across and penetrating 6 feet (1.8 m) or more into the ground. The entrance is often surrounded by a craterlike wall of coarse soil particles. The ants forage in large groups and the tracks leading to the nest look like well-marked roads. Within the nest there are special chambers, or granaries, in which the grain is stored.

The seeds are collected from living grasses and from standing crops of grain, or from the ground where they might have scattered during harvesting. In the Mediterranean and the Near East harvesting ants collect wheat, in Asia they take millet. In some regions these ants take so much of a cultivated crop that they have become serious grain pests.

The division of labor

The small workers gather the seeds and bring them home, where the husk or chaff is removed by the larger workers using their powerful jaws. These large individuals also crush the seeds when they are needed for food. Crushed and masticated grain is given to the growing larvae, and the adult ants also live on the store during times of drought when the foragers can find nothing edible outside the nest. The discarded

HARVESTING ANTS

PHYLUM	**Arthropoda**
CLASS	**Insecta**
ORDER	**Hymenoptera**
FAMILY	**Formicidae**
GENUS	***Messor; Monomorium; Pheidole; Pogonomyrmex; Ischnomyrmex; others***
SPECIES	**Many**

ALTERNATIVE NAME
Harvester ant

LENGTH
***Messor* worker: up to ⅖ in. (1 cm)**

DISTINCTIVE FEATURES
Highly polymorphic (appearing in various forms). Larger workers: proportionately huge heads; powerful mandibles.

DIET
Harvested seeds and grains

BREEDING
After mating, new queen feeds first batch of young with saliva; these workers tend subsequent eggs; females come from fertilized eggs; males from unfertilized eggs

LIFE SPAN
Queen: several years. Worker: less than this.

HABITAT
Deserts and dry grasslands

DISTRIBUTION
Warm regions of Mediterranean, Africa, Asia, North and South America and Australia

STATUS
Common

Accidental crops

When the ants fail to prevent germination, they bring the sprouting grain to the surface and throw it on the rubbish heap, where some of the seeds may germinate successfully and produce plants. It was once thought that these ants deliberately cultivated such crops around their nests.

Some plants are specially adapted to be dispersed by harvesting ants. The seeds have food bodies called elaiosomes and special chemical attractants that stimulate the ants to collect them. When the ants grip the elaiosome they take the whole seed. The elaiosome is removed, fed to the developing larvae in the nest and the hard seed is discarded, intact and viable. This phenomenon is termed myrmecochory and is found worldwide.

New colonies

The life cycle of the harvesters is similar to that of other ants. Winged females mate with winged males. After mating, the males die and the females shed their wings. Each female or queen then starts a new colony, seeking out a crevice in which to lay her eggs and raise a small brood. The queen feeds this first batch of young with saliva and these become workers. Eggs continue to be laid by the queen or queens and further larvae are tended and fed by the workers. The young pass through a pupal stage before reaching maturity. Females are raised from fertilized eggs, males from unfertilized eggs. Whether a female becomes a worker or a reproducing female depends on diet.

Harvesting ants (genus Messor, above) take grass seeds to the nest. These ants also harvest millet and grain, and in some regions have become a serious pest to cultivated crops.

chaff is thrown outside on a rubbish heap, which comes to form a ring around the nest and is a clearly visible feature of a flourishing colony.

Well-kept granaries

Within the nest the storage chambers are kept well drained so the seed remains dry and does not germinate. If the seeds begin to germinate the ants bite off the embryonic root or radicle, preventing further growth. When heavy rain does penetrate to the storage chambers, the damp seed is brought to the surface and spread out around the nest to dry. The quality of the seed as food is improved by this treatment, as the starch in the seeds is partly converted to sugar.

HARVEST MOUSE

An Old World harvest mouse among oxeye daisies. Harvest mice have a 3-hourly rhythm of alternately eating and sleeping. For every ½ hour spent feeding, they sleep for 2½ hours.

EXCEPT FOR THE PYGMY SHREW the Old World harvest mouse is the smallest mammal in Britain. It also ranges across Europe, except for the Mediterranean region, and east across Siberia and China to Taiwan. Another group also called harvest mice, the American harvest mice, is found across western Canada, North and Central America, Colombia and Ecuador. The American harvest mice belong to a different family but were given the same name because they resemble the Old World species in their habits.

The Old World harvest mouse is about 5 inches (13 cm) long, of which nearly half is a scaly, almost naked tail. It weighs no more than ¼ ounce (7 g). Its fur is soft and thick, yellowish red or russet with white underparts. It has a rather volelike head with a blunt nose, black, medium-sized eyes and large, rounded ears. The tail is prehensile on its outer part. The Old World harvest mouse belongs to the family Muridae and there is just one species, *Micromys minutus*.

American harvest mice

There are 20 species of American harvest mice belonging to the family Cricetidae. These include the western harvest mouse (*Reithrodontomys megalotis*), the eastern harvest mouse (*R. humulis*) and the salt marsh harvest mouse (*R. raviventris*). The American species are often larger than the Old World harvest mouse, up to 6 inches (15 cm) of head and body with as much as 4½ inches (11.5 cm) of tail. Although they have certain habits in common, and all are good climbers, the American species do not have a prehensile tail, nor do they nest on or near the ground as the European species does. Rather their nests are normally lodged in a shrub well above ground.

Rhythm of eating and sleeping

The Old World species prefers tall grassland vegetation and is often found in the rank herbage of ditches, pastures and fields of cereal crops. It sometimes lives in salt marshes and reed beds.

HARVEST MICE

ORDER **Rodentia**

FAMILY **Muridae**

GENUS AND SPECIES **Old World harvest mouse,** *Micromys minutus*; **American harvest mice,** *Reithrodontomys* **(20 species)**

WEIGHT
Old World harvest mouse: ⅕–¼ oz. (5–7 g). American harvest mice: ⅕–¾ oz. (6–20 g).

LENGTH
Old World harvest mouse. Head and body: 2⅛–3 in. (5.5–7.5 cm); tail: 2–3 in. (5–7.5 cm). American harvest mice. Head and body: 2–6 in. (5–15 cm); tail: 1¾–4½ in. (4.5–11.5 cm).

DISTINCTIVE FEATURES
Old World harvest mouse: tiny size; russet coat; long, prehensile tail. American harvest mice: gray to brown back, with paler belly; large eyes and ears; long tail.

DIET
Cereal and grass seeds; invertebrates; shoots

BREEDING
Breeding season: varies, may be all year; number of young: usually 3 or 4; gestation period: 17–18 days (Old World harvest mouse), 20–25 days (American harvest mice); breeding interval: several litters per year

LIFE SPAN
Old World harvest mouse: a few months. American harvest mice: about 18 months.

HABITAT
Old World harvest mouse: tall grassland vegetation. American harvest mice: generalists.

DISTRIBUTION
Old World harvest mouse: western Europe east to China and Taiwan. American harvest mice: western Canada south to Ecuador.

STATUS
Old World harvest mouse: at low risk. Some American species threatened or endangered.

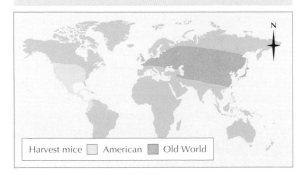

Harvest mice ▢ American ▮ Old World

All species of harvest mice are agile climbers among vegetation, often using the tail for balance.

During its waking periods the harvest mouse spends most of its time climbing about the stalks of cereals or other stout plants. It is known to have a 3-hourly rhythm of alternately feeding and sleeping throughout each 24 hours. Every third hour, night and day, it feeds for half an hour, and spends 1½–2 hours of every 3 hours asleep. It nests in long grass, its sleeping nest being a mass of grass blades shredded lengthwise.

Although it used to be common, the Old World harvest mouse is now near threatened across much of its range, due to modern agricultural methods such as the use of reaping machines and earlier harvesting.

The American species are generalists in terms of their habitat, and are often found on farmland, marshland and grassland. They some-

The Old World harvest mouse is distinguishable from its 20 American cousins in that it is usually smaller, has a prehensile tail and builds its nest near to the ground.

five toes on each hind foot is large and opposable to the rest. A harvest mouse can grip a stem with each hind foot and take hold with its tail, leaving the front paws free for feeding. It can also use the tail for balance. It will run up a stem that bends under its weight, swinging its tail from side to side, much as a tightrope walker uses a pole.

Choice of cereals

The Old World harvest mouse does not hibernate but winters in burrows in the ground. It will also tunnel into hayricks or cornricks, but these are going out of use in modern agriculture. All species of harvest mice feed on a variety of seeds, especially of grasses and cereal, along with green shoots. In summer they will also take a certain number of insects and other small invertebrates.

Nest woven from grasses

The breeding season for the Old World species is mainly from April to September. The female makes a round nest 3 inches (7.5 cm) in diameter, woven of grass or wheat blades split lengthwise and slung between two or three stalks. The nest has no definite entrance, the material being pushed aside for entrance or exit. The male is not allowed into the nest. After a gestation period of 17 or 18 days a litter of three to eight young is born. A female may have several litters in a season. The babies open their eyes at 8 days, make their first excursions from the nest at 11 days and are independent after 15 days. Later they take on the distinctive reddish tint, beginning at the hindquarters and gradually extending forward.

The American harvest mouse is known to breed throughout the year. Gestation is also longer in the American species, around 20 to 25 days. The female normally gives birth to fewer young than in the Old World species, three or four babies being the average. Young American harvest mice open their eyes at 5 days and leave the mother at 10 days.

The natural expectation of life for the Old World harvest mouse is up to 5 years in captivity, but much less in the wild, perhaps as little as a few months. American harvest mice live 3 or 4 years in captivity but survive longer than their Old World relatives in the wild, up to 18 months. Little is known of predators, but harvest mice are often killed by small carnivores as well as by raptors (birds of prey).

times also live in tropical and montane forests. The salt marsh harvest mouse is endangered as a result of habitat loss through development and marsh drainage. Other species are threatened.

Agile climbers

The most noticeable feature of all harvest mice is their agile climbing. Here the tail plays a large part. The moment the harvest mouse stops moving among vegetation, the end of its tail curls around a stalk. While the mouse is moving, the tail is constantly taking a partial grip, ready to take a firmer hold when necessary. The mouse can also grip with its hind feet. The outer of the

HATCHET FISH

Tiny, strangely shaped fish, looking like strips of shiny, crinkled tinfoil—such is the best description of the 42 species of deep-sea hatchet fish. Most species are 1–2½ inches (2.5–6.5 cm) long, the largest being no more than 5 inches (13 cm). There are 450 of the smaller ones to the pound (0.5 kg).

Hatchet fish have deep, high bodies flattened from side to side, resembling the head of a hatchet, the lower surface corresponding to the sharp edge of the hatchet blade. They are covered with large scales, which in a few species are missing from the breast and belly, leaving those parts transparent. In all species the color of the body is silvery and iridescent. Their eyes are large, and their fins are of moderate size and transparent except for the rays supporting them. Along the lower edge of the body and on the underside of the tail are many closely set light organs. The light from these is usually blue but in some species a bright ruby red or yellow light has been seen.

The marine hatchet fish should not be confused with the nine species of freshwater fish also given this name. They belong to the genera *Gasteropelecus*, *Thoracocharax* and *Carnegiella*, and are found in northern South America.

Sensitive telescopic eyes

Marine hatchet fish live in the twilight zone of the oceans, where only the green and blue rays of light penetrate. They can be found at up to 5,000 feet (1,525 m) in most tropical and temperate seas. The human eye can detect light at depths of about 1,500 feet (450 m), although sensitive photographic plates lowered into the sea register

Hatchet fish are named for their compressed bodies, which have a sharp and bladelike lower edge. These fish abound in the dimly lit depths of the world's seas and oceans.

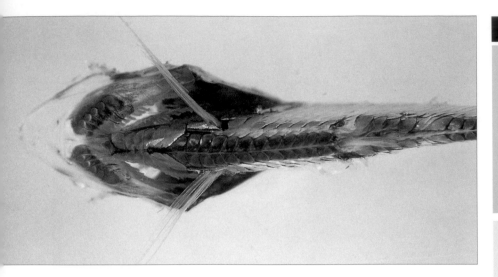

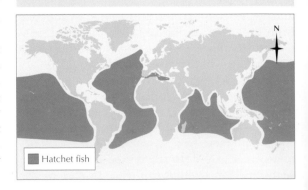

HATCHET FISH	
CLASS	**Osteichthyes**
ORDER	**Stomiiformes**
FAMILY	**Sternoptychidae**
GENUS	***Argyropelecus, Polyipnus* and *Sternoptyx***
SPECIES	**42, including *A. hemigymnus, A. gigas, A. aculeatus* and *S. diaphana***

LENGTH
***A. gigas*: up to 5 in. (13 cm); other species: 1–2½ in. (2.5–6.5 cm)**

DISTINCTIVE FEATURES
Deep, extremely compressed body tapering to sharp edge below belly; thin tail; mouth nearly vertical; large eyes, telescopic in some species; silvery and iridescent body with transparent fins; clusters of light organs on underside

DIET
Small species: tiny crustaceans such as copepods and ostracods; larger species: krill and salps (relatives of sea squirts)

BREEDING
Very poorly known

LIFE SPAN
Not known

HABITAT
***Argyropelecus*: deep seas at depth of 330–2,000 ft. (100–600 m). *Polyipnus*: coastal waters at depth of up to 1,300 ft. (400 m). *Sternoptx*: deep seas at depth of 1,650–5,000 ft. (500–1,525 m).**

DISTRIBUTION
***Argyropelecus* and *Sternoptx*: virtually worldwide in tropical, subtropical and temperate waters. *Polyipnus*: most species in western Pacific.**

STATUS
Abundant

A view of the lower surface of a preserved hatchet fish specimen. Most hatchet fish have clusters of light organs on the underside.

that a very small amount of light penetrates even farther, down to 3,000 feet (900 m). We can suppose that the large eye of hatchet fish, with its large lens and retina composed of long rods only, is at least as sensitive as the human eye.

How much the light from the hatchet fish's own light organs (which are on the lower edge of the body) help the eyes is difficult to determine. They probably help little, since the eyes are well up on top of the head or directed upward. In some species the eyes are tubular and are usually described as telescopic.

The light organs may function as a form of camouflage. For instance, if a predator is swimming underneath a hatchet fish, it will not be able to identify the fish because the emitted light is from the same spectrum and of the same intensity as that of the surrounding environment.

Submarine weightlessness

Hatchet fish are very light, weighing on average about ⅓₅ ounces (0.8 g). They have a well-developed swim bladder. These two things together mean the fish have neutral buoyancy. We are accustomed to the idea of weightlessness in space travel; neutral buoyancy means much the same thing. So hatchet fish can swim easily and make considerable vertical migrations daily, coming up almost to the surface at night and going down again by day. In these migrations they are following their food, which consists of small planktonic animals such as copepods and ostracods (types of crustaceans).

Sailors on U.S. Navy ships in the Pacific noted that their echo sounder traces showed, in addition to a profile of the seabed, a second, and sometimes a third or fourth, profile far above the seabed. These "deep scattering layers," as they came to be called, proved to be large planktonic animals: jellyfish, crustaceans and arrowworms. In the trace of a deep scattering layer are many blobs: hatchet fish and lantern fish.

Hatchet fish

HAWAIIAN GOOSE

NCE ALMOST EXTINCT in the wild, the Hawaiian goose, often known by its Hawaiian name of nene, has been saved by careful breeding in captivity. It is in fact a relative of the Canada geese, *Branta canadensis*, that settled on the Hawaiian islands thousands of years ago. The Hawaiian goose evolved into the nonaquatic, nonmigratory bird with half-webbed feet that it is today. It is a medium-sized goose, males weighing about 4¾ pounds (2.2 kg) while females weigh some 4¼ pounds (1.9 kg). It measures 1⅞–2⅓ feet (56–70 cm) from head to tail. The forehead, crown and chin, and a wide band down the back of the neck, are black. The sides of its head and the rest of the neck are tawny yellow with dark, diagonal stripes running down the neck. The rest of its body is grayish brown with whitish buff barring, except for a pale brown breast. The vent area and uppertail coverts are white. The female is smaller than the male.

Danger from lava

The Hawaiian goose was almost wiped out before studies could be made of its behavior in the wild. Hawaiian geese are also called "lava geese" as they are found among the lava fields near Mauna Loa, a very active volcano on the island of Hawaii. It is quite possible that the wild population in this region could be wiped out by one flow of lava if caught during the flightless period when the geese molt their primary wing feathers. These geese are also found on the island of Maui northwest of Hawaii.

It appears that at one time the greater part of the Hawaiian goose population lived in flocks in the hills of their natural habitat and came down to lower ground to breed. In the uplands the geese lived in dry country among the lava at altitudes of 5,000 to 8,000 feet (1,500–2,500 m). This is no doubt the explanation for the strength of the Hawaiian goose's legs and feet and the small size of the webs between its toes, which are half the size of those of other geese.

Aggressive males

Hawaiian geese feed on succulent leaves, stems, flower buds and berries, and will also strip seeds from grasses. In September the flocks that have been feeding together split up and each pair forms a territory that is defended against other Hawaiian geese. The ganders (males) can be extremely aggressive and in captivity pairs have to be isolated from one another. Not only are the territory and the brood defended vigorously against all comers, but the ganders will also sometimes attack their own mates.

Winter breeding

The nest is built in a hollow in the ground where some three to five eggs are laid, generally on alternate days. These are incubated by the goose (female) while the gander stands guard. The goslings, which hatch after an incubation period of about 29 days, are brown with whitish markings. They run about on large feet under the protection of their parents, feeding on plants such as watercress and sow thistle. Their feathers do not appear for 5 weeks, and the young geese

A Hawaiian goose or nene in its natural habitat, still an uncommon sight despite regular reintroductions of the species to Hawaii, Maui and Kauai.

HAWAIIAN GOOSE

CLASS	**Aves**
ORDER	**Anseriformes**
FAMILY	**Anatidae**
GENUS AND SPECIES	***Branta sandvicensis***

ALTERNATIVE NAMES
Nene; lava goose

WEIGHT
**Male: average 4¾ lb. (2.2 kg);
female: average 4¼ lb. (1.9 kg)**

LENGTH
Head to tail: 1⅘–2⅓ ft. (56–70 cm)

DISTINCTIVE FEATURES
**Medium-sized goose with half-webbed feet
and rather short wings. Black forehead,
crown, chin and back of neck; tawny yellow
sides of head and neck; dark diagonal
markings on neck; gray brown body with
buff-white barring and paler breast.**

DIET
**Grasses, herbs, leaves, buds, berries
and seeds**

BREEDING
**Age at first breeding: 2–3 years; breeding
season: winter months (October to January);
number of eggs: 3 to 5; incubation period:
about 29 days; breeding interval: 2 or 3
broods per year**

LIFE SPAN
Not known

HABITAT
**Lava flows on poorly vegetated volcanic
slopes at 5,000–8,000 ft. (1,500–2,500 m)**

DISTRIBUTION
**Hawaii and Maui islands; Kauai island
(reintroduced)**

STATUS
Vulnerable

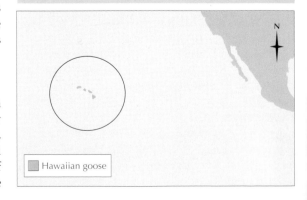

Hawaiian goose

*Hawaiian geese (family
of captive birds, above)
are unusual in that they
breed during the winter
months when the days
are short. Development
of the young geese is
slow because they
have less time to feed.*

cannot fly until they are 3 months old. By
contrast, Canada geese can fly at half this age.
They are able to grow much more rapidly
because summer days are longer in northern latitudes, so there is more time for feeding. The
Hawaiian geese also breed from October to
January, the time of year when the days are
shortest. This is a most unusual time for animals
to breed and no reason is known for it. The goose
starts to breed when 2 or 3 years old and lays
two or three clutches of eggs each year.

Wrong close season

Because no one studied the habits of Hawaiian
geese, it was not known for a long time that they
bred in the "wrong" half of the year. As a result,
the open season for shooting them was fixed
during their breeding season. This was one of
several reasons for their rapid decline. They were

never abundant, and the total population when Europeans arrived in Hawaii was probably about 25,000 individuals. Dogs, pigs, rats and cats were then introduced to the island and preyed on the geese. However, their decline was particularly marked from when mongooses were liberated on the island at the end of the 18th century.

Saving the nene

From an estimated 25,000, the population of the Hawaiian goose had sunk to just 34 by 1950. Of these, 17 were held in captivity and 17 were still alive in their natural home. Efforts were then made to preserve the species. The wild ones were protected by banning hunting altogether. Stocks of captive ones were increased by breeding, notably at the Severn Wildfowl Trust at Slimbridge, England. In captivity it was found that the geese were liable to desert their eggs, which had to be hatched under bantams, and some goslings had to be helped out of their shells. The hatching rate for Hawaiian geese is considerably lower than that of other geese in the Slimbridge collection. This may be another reason for the small size of the original population and the rapid decline in their numbers.

By 1957 the numbers had risen to 129. Of these at least 35 were wild and 49 of the captives were living outside Hawaii, 40 of them at Slimbridge. The Slimbridge flock grew apace and

in 1962 there were enough to fly 32 back to the Pacific and release them on Maui, with more being returned in the next 2 years. In 1964 the world population of Hawaiian geese had risen to between 380 and 420 birds, of which 250 to 300 were wild or captive on Hawaii and Maui. By 1970 the world population had risen to more than 900, over half of which were living wild.

The Hawaiian goose is lucky because the keeping of wildfowl is a popular pursuit. Its plight came to the notice of people who were committed to saving it and were already in a position to keep a small stock and breed them.

Problems with conservation

In the period between 1960 and 1990 more than 2,100 birds were released on the islands of Hawaii and Maui, and 12 were released on Kauai nearby. Of course not all of these survived. In 1990 the populations on the three islands were up to 339, 184 and 32, respectively. Potential problems with their continued survival today include interbreeding, the loss of adaptive skills, disease, poaching, road kills, dietary deficiencies and, still, predation by cats, dogs and mongooses.

Of the eight areas where the geese were released, one has a stable population and two populations are increasing due to intensive control of predators. The remaining populations are dependent on continued reintroductions.

A captive Hawaiian goose. Most wild populations depend on captive breeding to maintain their numbers. The wild geese are still preyed upon by cats, dogs and mongooses, as well as being subject to both disease and poaching.

HAWKMOTH

AWKMOTHS, ALSO CALLED sphinx moths, make up the family Sphingidae. They are large, thick-bodied moths found all over the world, although most of the 900 species are tropical. Hawkmoths have long, narrow forewings and hind wings that are often shorter and almost triangular in shape. They have large eyes and the tongue or proboscis is well developed and sometimes extremely long. Hawkmoth larvae are stout and usually have a "horn" at the hind end

An elephant hawkmoth flying to a night-scented white campion flower. Most adult moths feed on nectar, hovering and probing the flowers with their long tongue, or proboscis.

Hover to drink nectar

The hawkmoths all have powerful and well-controlled flight, and it has been found that some species can hover like hummingbirds when feeding from flowers. The small hummingbird hawk, *Macroglossum stellatarum*, is one of these species, and it hovers in front of flowers such as honeysuckle, jasmine and valerian. Some of the larger species are very swift in flight, but since almost all of them fly by night their speeds are extremely difficult to measure. However, speeds of 33 miles per hour (53 km/h) have been recorded in the past.

If the moth is disturbed shortly after flight, it is still likely to be at a sufficient temperature to enable it to fly away. A hawkmoth that has been at rest for some time, on the other hand, often cannot fly immediately because the temperature of its body is too low. If such a moth is disturbed, it will rapidly vibrate its wings for a minute or more before taking to the air. This raises its internal temperature to the point at which it can fly.

Migratory habits

As might be expected, the hawkmoths' powerful flight is associated with their migratory habits. The convolvulus hawkmoth, *Agrius convolvuli*, is a huge gray moth with a very wide range, from Africa and Europe east across Asia to Australia. It breeds in Africa and regularly flies northward, crossing the Alps with ease and appearing in Britain in late summer and autumn. It has even been recorded as far north as Iceland. The equally large but much more heavily built death's head hawkmoth, *Acherontia atropos*, is seen less frequently in Britain but occasionally migrates from North Africa in swarms.

Variety of food plants

Hawkmoth caterpillars feed on the leaves of plants and trees. The various species usually confine their attention to a fairly narrow range of related plants. Both their scientific and English names often indicate the food plant, as in the poplar hawkmoth (*Laothoe populi*), the lime hawkmoth (*Mimas tiliae*) and the privet hawkmoth (*Sphinx ligustri*).

The caterpillar of the death's head hawkmoth can be huge. When fully fed and developed it can reach 5 inches (13 cm) in length. The eggs are laid singly on the upper surface of the host plant. The death's head larva feeds on plants of the family Solanaceae, which

DEATH'S HEAD HAWKMOTH

PHYLUM	**Arthropoda**
CLASS	**Insecta**
ORDER	**Lepidotera**
FAMILY	**Sphingidae**
GENUS AND SPECIES	***Acherontia atropos***

LENGTH
Adult wingspan: 4–5½ in. (10–14 cm).
Larva: up to 5 in. (13 cm).

DISTINCTIVE FEATURES
Adult: large, heavily built body; large eyes; brown and yellow coloration; distinctive pale skull marking at rear of thorax.
Larva (caterpillar): very large size; usually yellowish in color, occasionally brownish or blackish.

DIET
Adult: occasionally tree sap; also honey from beehives and wild bees' nests.
Larva: plants of Solanaceae family, such as potatoes, tomatoes, eggplants, nightshades and jasmine.

BREEDING
Number of eggs: small numbers, laid singly; breeding interval: depends on environmental conditions such as temperature

LIFE SPAN
Larva: up to 1 year

HABITAT
In and around host plants

DISTRIBUTION
Throughout warmer regions of Europe, Middle East and Africa; migratory in some parts of range

STATUS
Uncommon

includes many cultivated crops such as potatoes, sweet peppers and eggplants. More toxic members of this family of plants include the deadly nightshade and the woody nightshade.

The caterpillars of many hawkmoths are big enough to look rather like snakes. Some tropical species have false eyes and other markings on the front part of the body and these, with appropriate movements and postures, give the caterpillars a snakelike appearance. It is thought that this might be effective in scaring off hungry birds. The conspicuous "eyes" on the fore part of the elephant hawkmoth caterpillar, *Deilephila elpenor*, might serve a similar purpose in deterring would-be predators.

Pupate underground

Hawkmoths lay their eggs on the host plants. Most of the hawkmoth caterpillars pupate underground in an earthen cell, but a few, including the two elephant hawks, spin an open-mesh cocoon among debris or under herbage on the surface of the ground. The pupae of these species may also be found just under the soil's surface. In those species in which the proboscis is greatly developed, it is contained during the pupal stage in a curved or coiled sheath at the front of the pupa. This is present in the pupa of the privet hawkmoth and is very conspicuous in that of the convolvulus hawkmoth.

Adapted to suck nectar

Almost all adult hawkmoths feed on nectar, hovering and probing the flowers with the long, tubular proboscis. This is coiled when not in use.

An oleander hawkmoth, Daphnis nerii. *Many hawkmoths have distinctive markings that look like the eyes of larger animals or might give the moths a snakelike appearance. Such protective mimicry is thought to deter predators.*

that a hawkmoth would be found that could do this. In 1903 just such a moth was discovered in Madagascar with a proboscis 11 inches (28 cm) long. It was named *Xanthopan morgani* and received the appropriate subspecific name *predicta*.

Robbing beehives

The death's head hawkmoth has a short, stiff proboscis, quite unsuitable for delicately probing flowers. It will sometimes feed on tree sap but the adult moths have also been known to rob bees' nests and hives of their honey. Most modern hives are designed to prevent entry by this moth, but the beekeepers of a century ago knew it well, and some early entomologists called it the "bee tyger."

The death's head hawkmoth's modern name is based on a rather fanciful resemblance in the markings on its thorax to a human skull. Another curious feature is its ability to squeak loudly, both as a caterpillar and as a moth. The caterpillar stridulates, while the adult moth squeaks by forcing air out through an opening at the base of the proboscis. In the past it was suggested that the adult moth squeaked to pacify the bees as it took their honey. It is now thought that the squeak actually functions to frighten off predators.

Fake eye trick

The eyed hawkmoth, *Smerinthus ocellata*, is one of the most attractive of hawkmoth species. The elegantly shaped forewings are marbled in shades of violet gray, and on each of the pink and ocherous hind wings is a sharply drawn, blue-and-black "eye." However, an eyed hawkmoth sitting at rest, with its hind wings covered, looks more like a pile of dead leaves. If discovered and pecked by a bird, it reacts by lifting its forewings and revealing what appear to the predator to be a pair of lurid, staring eyes.

Experiments have been carried out in which birds were observed attacking an eyed hawkmoth. In each case the bird started back in alarm at the moth's defensive behavior. Some bolder birds returned to the attack, but some were wholly daunted and left the moth alone. We can take it that the lives of some eyed hawkmoths are saved by this device.

The two bee hawkmoths of the genus *Hemaris* (*H. tityus* and *H. fuciformis*) have partly scaleless and transparent wings. This makes them look very much like bumblebees and so these hawkmoths may also gain some protection from their appearance.

A caterpillar of the death's head hawkmoth in a defensive posture. This species is also able to stridulate, producing a high-pitched squeak, in an attempt to ward off predators.

That of the convolvulus hawk is extremely long, around 4 inches (10 cm) when fully extended, and can take nectar from flowers, such as those of tobacco plants, with very long corolla tubes. This species has, however, by no means the longest tongue of all hawkmoths. In 1862 the English naturalist Charles Darwin noticed that there was an orchid native to Madagascar, with its nectaries situated at a depth of 10–13 inches (25–33 cm). No insect was known that could reach them and so act as a pollinator. It was predicted

HAWKSBILL TURTLE

THE HAWKSBILL TURTLE IS the source of tortoiseshell, once widely used for combs and eyeglass frames as well as for ornamental items. The introduction of plastics at first destroyed the market for tortoiseshell, but no plastic has ever rivaled this material for its translucent coloring. It is now being used again, posing a threat to the survival of this species.

One of the smallest turtles, the hawksbill usually has a shell around 2 feet (60 cm) long and weighs about 100 pounds (46 kg). The largest hawksbill found weighed 280 pounds (130 kg). The shell is largely brown on top, often with dark blotches at the front. The plastron, the underpart of the shell, is yellow. The name hawksbill is derived from the hooked beak, which is far more prominent than the snouts of other turtles.

Hawksbills are found in tropical seas around the world. On the Pacific side of America they are found as far north as Oregon and as far south as Peru. On the Atlantic side, the warm Gulf Stream allows them to spread farther north. They are known to breed in Florida, and they occasionally stray as far north as Massachusetts and even to the Orkney Islands on the eastern side of the Atlantic. To the south hawksbills reach southern Brazil, and, although rare on the western coast of Africa, they are found around Madagascar and in many other parts of the Indian Ocean.

Home-loving turtles

Unlike green or ridley turtles, hawksbills are found only in scattered numbers on their quiet breeding and feeding grounds. Every secluded beach is visited by a few female hawksbills in search of egg-laying sites, and some hawksbills are almost certain to be found feeding around any coral reef or coral rock. Hawksbills do not migrate regularly but appear to stay more or less in one place, although some do wander long distances. They are slow swimmers compared with green turtles, and often have large barnacles on their shells.

The hawksbill has never been as important a source of human food as the green turtle, so its habits have not been as well studied. Although they eat water plants, these turtles prefer animal food such as crabs, fish and planktonic animals, cracking open crabs and shellfish with their powerful beaks. This results in their flesh being less acceptable to humans, because it often has a strong, rather oily taste. It is also sometimes considered toxic, and in some places fishers test for poison by first throwing a hawksbill's liver to the crows. This belief that the flesh is poisonous might be due to the hawksbill's habit of eating jellyfish. It attacks them with its eyes closed for protection, but must be stung by the nematocysts (stinging cells) in its mouth and throat as it swallows the jellyfish. It is probable that, like other animals that feed on poisonous creatures, hawksbills are themselves totally or partially immune to their prey's poison.

Long breeding season

As well as breeding at well-spaced intervals, hawksbills lay their eggs at almost any time of the year. Peak laying takes place in May and June in the Caribbean and from September to November in the Indian Ocean. Hawksbills do not congregate in such vast numbers as some other species of marine turtles, but they do come ashore on breeding beaches. They sometimes share beaches with green turtles, although they lay their eggs at a different time of year. Hawksbill breeding beaches are found around the Pacific, Indian and Atlantic Oceans and in the Caribbean, Mediterranean and Red Seas.

Each adult female hawksbill lays eggs probably every 2 or 3 years. Each clutch contains, on average, 160 rounded eggs, each about 1½ inches (3.7 cm) in diameter. As a female crawls up the

Hawksbill turtles are famed for being the source of tortoiseshell, one of the reasons they are now endangered.

HAWKSBILL TURTLE

CLASS	**Reptilia**
ORDER	**Testudines**
FAMILY	**Cheloniidae**
GENUS AND SPECIES	***Eretmochelys imbricata***

WEIGHT
Average 100 lb. (46 kg)

LENGTH
Usually about 2 ft. (60 cm); larger specimens up to 3 ft. (90 cm)

DISTINCTIVE FEATURES
Narrow, elongated snout, rather like a hawk's bill; scutes (plates) overlap in young and fit smoothly in adults; generally brownish above, yellowish below; elaborate patterns on carapace (shell), but may fade with age

DIET
Invertebrates such as crabs, sea urchins, squid and jellyfish; also corals, sponges, small fish and plant material

BREEDING
Age at first breeding: probably 20 years or more; breeding season: all year with peak in May–June (Caribbean) or September–November (Indian Ocean); number of eggs: about 160; hatching period: 50 days; breeding interval: 2–3 years

LIFE SPAN
Probably up to at least 45 years

HABITAT
Mainly coral reefs and rocky areas in shallow waters

DISTRIBUTION
Open seas almost anywhere between 25° N and 25° S; sometimes strays farther, especially in Northern Hemisphere

STATUS
Endangered

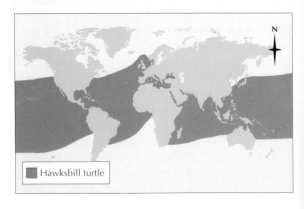

Hawksbill turtle

Studies on tagged hawksbills have shown that they do not reach sexual maturity until at least 20 years of age. They may then live for a further 25 years or even longer.

beach she nuzzles the sand at intervals, apparently searching for a suitable site for the nest. She does not dig a body pit like the green turtle, but only an egg hole about 10 inches (25 cm) deep. As in all marine turtles, the temperature at which an egg develops determines the sex of the young that hatches from it. Below 82° F (28° C) all the young are males, while above 90° F (32° C) all are females. The eggs hatch in 50 days, and the young turtles have the same hazardous dash to the sea as other baby turtles.

Tortoiseshell plates

The plates of this turtle's shell have been used for hundreds of years. Since early times, the Chinese and Japanese used tortoiseshell in works of art and the Romans used it as a veneer for furniture. It was, and still is, popular because of its translucent or clear amber coloring flecked with black, red, green or white. The advent of synthetic plastics appeared to have saved the hawksbill from possible extinction. Then the demand rose again, as no plastic could rival real tortoiseshell for its beauty. Tortoiseshell souvenirs also became popular as the postwar tourism boom developed.

Hawksbill turtles are now classified as endangered and there are international laws banning the trade in tortoiseshell. Despite this, tortoiseshell souvenirs continue to be sold, and the turtles face other threats to their survival. These include disturbance of nesting beaches, having their eggs collected, the hunting of adults for food and as tourist curios, and drowning in fishing nets.

HEDGEHOG

THE HEDGEHOG OR HEDGEPIG is an animal remarkable for its coat of spines, made from thickened hairs, and its habit of rolling into a defensive ball when frightened.

The male, or boar, Eurasian hedgehog grows to 1 foot (30 cm) long, with ⅜–2 inches (1–5 cm) of tail. The sow (female) is on average ¼ inch (2 cm) shorter. Both sexes weigh up to 2⅖ pounds (1.1 kg). A hedgehog's neck and body are short in relation to its bulk. The back and top of its head are coated with sharply pointed spines, ¾ inch (2 cm) long, with each spine set at an angle to the skin. The rest of the body is covered in coarse hair. All four feet have five clawed toes and five pads on the sole.

Besides the Eurasian hedgehog, *Erinaceus europaeus*, so called because it ranges across Asia as well as Europe, there are 11 other species in southwestern Europe, Asia and Africa. All are alike in habits and differ only slightly in appearance.

Rolls into a ball for defense

The Eurasian hedgehog lives in a variety of habitats, wherever there is enough food and enough vegetation cover or dry leaf litter into which it can retire to sleep through the day. Such habitats include forest, farmland, scrub and suburban parks and gardens. The foot of a hedge is a typical place to find the animal asleep. At twilight it comes out to forage. When undisturbed, it moves quickly over the ground, despite its short legs. At any unfamiliar sound or movement it stops dead, drawing the spines forward on the top of its head. This is in preparation for the hedgehog rolling itself up if its alarm increases. When it does so, its head and legs are withdrawn and the edges of the prickly mantle are drawn around them to present an almost complete ball of spines. A hedgehog climbs and swims well, but usually keeps to the ground. Its sight seems to be poor, but its senses of smell and hearing are acute.

Twilight forager

Snails, slugs, insects and earthworms form the normal hedgehog diet. Small vertebrates such as mice and rats may also be taken at times, along with frogs, lizards and snakes. Although hedgehogs were thought to eat no plant material, they are actually omnivorous, feeding on acorns and berries, and tame hedgehogs have been known to eat fruit. Some hedgehogs take the eggs of ground-nesting birds, including quails, partridges and pheasants.

One legend of the past is that hedgehogs will actually take the milk from cows, deliberately lactating the teats to do so. However, there is no evidence to support this belief. Nonetheless, for centuries hedgehogs had a bounty on their heads in England for this reason, and churchwardens' accounts used to record sums of money paid out annually for the slaughter of hedgehogs.

Although the Eurasian hedgehog is usually thought of as a woodland animal, hedgehogs in fact live in a variety of habitats. Some species, like this long-eared hedgehog, Hemiechinus auritus, are found in arid scrub, steppe and semidesert.

Rubbery-spined babies

The breeding season for the Eurasian species is in spring and summer, usually between May and July, but there may be a second litter during August and September. After a gestation period of 30–35 days, a litter of 2 to 7 blind, deaf and helpless young is born. Each is up to 3¼ inches (8.3 cm) long and weighs ⅓–⅔ ounces (9.5–19 g). They are sparsely covered with pale, flexible spines.

Between 36 and 60 hours after birth a second coat of darker spines appears among the first spines, but the young hedgehog is unable to roll up until it is 11 days old. About 3 days later its eyes open, first one and then the other, over a period of 3 days. At this time a third set of spines begins to grow through, each spine ringed with a dark band in the middle and a light band on either side. The first two sets of spines are shed when the young hedgehog is a month old. By this time it has started to make short journeys from the nest and is being weaned. Its weight doubles in 7 days from birth and is increased tenfold by the age of 7 weeks. However, sexual maturity is not reached until the following year. The mother alone looks after the youngsters, the the male hedgehog taking no part.

A young Eurasian hedgehog unrolling. From 11 days old, hedgehogs are able to roll themselves into almost a complete ball of spines whenever they feel threatened.

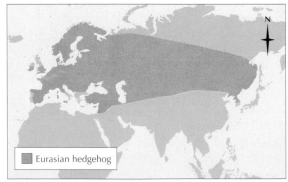

EURASIAN HEDGEHOG

CLASS	**Mammalia**
ORDER	**Insectivora**
FAMILY	**Erinaceidae**
GENUS AND SPECIES	***Erinaceus europaeus***

ALTERNATIVE NAMES
Hedgepig; urchin (archaic)

WEIGHT
1–2⅖ lb. (0.5–1.1 kg)

LENGTH
Head and body: 5⅓–12 in. (13.5–30 cm); tail: ⅖–2 in. (1–5 cm)

DISTINCTIVE FEATURES
Spikey coat on back, made of long bristles (thickened hairs); shorter, coarse coat on head and belly; pointed snout; rolls into a ball when frightened

DIET
Mainly invertebrates such as snails, slugs, beetles and earthworms, supplemented with small vertebrates and bird eggs; also nuts and berries

BREEDING
Age at first breeding: 10–11 months; breeding season: spring and summer; number of young: 2 to 7; gestation period: 30–35 days; breeding interval: 1 or 2 litters per year

LIFE SPAN
Up to 7 years

HABITAT
Woodland, hedgerows, farmland, scrub and more open habitats; also in suburban areas

DISTRIBUTION
Northwestern Europe east to East Asia; introduced to New Zealand

STATUS
Locally common

☐ Eurasian hedgehog

Eurasian hedgehogs live for a maximum of around 7 years, both in captivity and in the wild. Red foxes, *Vulpes vulpes*, and Eurasian badgers, *Meles meles*, are their principal predators apart from humans.

Intermittent winter sleep

In Eurasian hedgehogs hibernation covers the period from October to late March or April. Some individuals sleep through this whole period, but in others sleep is intermittent until December or even later. It is not unusual to see a hedgehog out and about on frosty nights, or even in snow, until the end of the year. It seems likely that sleep at the beginning of the winter is less profound for younger hedgehogs.

Before hibernating, hedgehogs choose a hole in a bank, perhaps one that has been enlarged by a colony of wasps. More commonly they choose a cavity between the buttress roots of a well-grown tree or under a heap of leaf litter, a favorite place being a compost heap. The nest is lined with dry leaves and moss, carried in the mouth, and the process appears to be started toward the end of summer. During hibernation the body is nourished by fat accumulated during the summer. However, energy requirements are low because the temperature of the body drops, breathing is so slight that it can hardly be detected, and the pulse rate also drops considerably.

The body of a hedgehog contains special fat cells, referred to as brown fat. These cells are larger than normal white fat cells and contain more fat droplets; they release heat 20 times faster. In addition, their heat production increases rapidly as the temperature of the surrounding air drops, so this dark-colored fat acts like a thermostatically controlled electric blanket.

Despite the profound sleep of hibernation, the hedgehog remains physiologically sensitive to its surroundings. Should the temperature drop too low, the heart, which remains warmer than the tissues on the outside of the body, automatically begins to beat faster. The animal will resume its temperature control and once more becomes warm-blooded. It will then take up its normal activity for a while, afterward falling asleep and continuing its hibernation.

Self-anointing

A strange habit of hedgehogs is that of self-anointing. The animal, discovering an unfamiliar substance, will lick it repeatedly while its mouth becomes filled with a frothy saliva. Whatever the substance, once the mouth is full of foam, the hedgehog raises itself on its front legs and throws its head first to one side and then to the other, placing flecks of foam on the spines with its tongue. There is as yet no fully satisfactory explanation for this behavior.

Hedgehogs hibernate in winter, but often their sleep is not continuous. Should temperatures drop too much they sometimes wake and resume their normal activity for a time before returning to hibernation.

HELLBENDER

THE HELLBENDER IS A giant salamander of North America, one of three species of the family Cryptobranchidae. The other two giant salamanders, genus *Andrias*, live in China and Japan. The origin of the name is not known, although it is thought to be connected with this amphibian's dark, almost sinister appearance and pliable body.

The hellbender can be up to 2½ feet (75 cm) long, its coloration ranging from light to dark brown to nearly black with scattered, blackish spots. Its body and head are flattened. The head is broad with a wide mouth and small eyes. Its tail is stout at the base, flattened and oarlike at the end, and accounts for around one-third of the animal's total length. Its skin is slimy and thrown into fleshy, dermal folds along the flanks. Its legs are moderately well developed, with four toes on the front feet and five on each hind foot. The hellbender lives in states such as Missouri, Indiana and Kentucky in the eastern United States.

The Chinese giant salamander resembles the hellbender in appearance but can be up to 5 feet (1.5 m) long. It is also much heavier, with beadlike barbels on its chin. The Japanese giant salamander is up to 4¾ feet (1.4 m) long, and can weigh up to 90 pounds (40 kg).

The Chinese giant salamander resembles the North American hellbender but at up to 5 feet (1.5 m) long, it is even larger and heavier.

Different ways of breathing

Hellbenders live in rivers and large streams with a rapid flow of water and, usually, a rocky bed. They swim by wriggling the body and waggling the head from side to side, an unusual method for a swimming animal. They lie up under large stones or among rocks by day, coming out at night to feed. Hellbenders breathe by internal gills, with two gill openings on either side of the neck. These are in addition to their lungs. When living in the still water of an aquarium, hellbenders are seen to rise to the surface every 25 to 40 minutes to take air. In well-aerated, running water, they surface less often. Breathing is also carried out through the skin, the folds of which are richly supplied with surface blood vessels.

The hellbender is able to bite hard when handled and has a reputation for being poisonous. It is, however, nonvenomous.

Generalist feeders

Hellbenders eat almost any aquatic animal small enough to be swallowed, including worms, water snails, insect larvae and crayfish, as well as frogs and fish, all captured with a quick sideways snatch. However, they do not feed when the temperature of the water falls below 50° F (10° C). Otherwise they feed well, scavenging dead animal flesh as well as taking live prey.

Some parental care

In August the male hellbender begins to remove silt from the downstream side of a rock, creating a bowl in which the female will later lay her eggs. These are ¼ inch (6 mm) in diameter and are laid in paired, rosarylike strings in depressions beneath the stones in the stream. The jelly around each egg swells to the size of a grape. Each jelly garland contains around 450 eggs and these are fertilized externally, as they are being laid. Several female hellbenders may use the same nest site. In both the Chinese giant salamander and the Japanese giant salamander, parental care of the eggs occurs.

The eggs hatch after about 120 days, and the larvae are a little over 1 inch (2.5 cm) long when they finally burst out of the jelly envelope. The larvae have tail fins and short gills. Their hind legs are at first paddleshaped, with no sign of toes, whereas

GIANT SALAMANDERS

CLASS	**Amphibia**
ORDER	**Caudata**
FAMILY	**Cryptobranchidae**
GENUS AND SPECIES	**Hellbender, *Crypto-branchus alleganiensis*; Chinese giant salamander, *Andrias davidianus*; Japanese giant salamander, *A. japonicus***

ALTERNATIVE NAME
Hellbender: hellbender salamander

WEIGHT
Japanese giant salamander: up to 90 lb. (40 kg)

LENGTH
**Hellbender: up to 2½ ft. (75 cm);
Chinese giant salamander: up to 5 ft. (1.5 m);
Japanese giant salamander: up to 4¾ ft. (1.4 m)**

DISTINCTIVE FEATURES
Huge size; flattened head and body; broad, powerful jaws; small eyes without eyelids; fleshy skin folds on sides of body; oarlike tail with stout base; brownish to black in color

DIET
Range of aquatic animals; also carrion

BREEDING
Age at first breeding: 5 years or more; breeding season: varies according to species; number of eggs: 450; incubation period: 120 days; breeding interval: not known

LIFE SPAN
Up to 30 years

HABITAT
Rivers and large streams

DISTRIBUTION
**Hellbender: eastern U.S., south to Arkansas.
Chinese giant salamander: central China.
Japanese giant salamander: Japan.**

STATUS
Not known

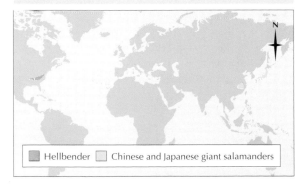

Hellbender ☐ Chinese and Japanese giant salamanders

the front legs show signs of two toes. The larvae swim with a flattened, rudderlike tail and, once they have eaten the remains of the egg yolk, they alternately swim against the current and probe the mud at the bottom of the river with the snout, searching for worms and other small aquatic animals. The external gills are absorbed when the larvae are 4–5 inches (10–13 cm) long and 1–1½ years old.

Hellbenders do not mature until they are at least 5 years old, and are neotenic: they achieve reproductive maturity while retaining larval external features. They have a long life span, estimated to be at least 30 years in the wild. One Japanese giant salamander lived 52 years in captivity, while the record for a captive hellbender is 55 years.

Predators of these amphibians are mainly freshwater turtles and large predatory fish. The Japanese giant salamander is fished with a line and hook and used for food.

Swamps and giants

The distribution of the giant salamanders, with two in northern Asia and one in North America, might seem strange. However this geographical separation is a familiar pattern for groups of animals that are of ancient lineage and are now dying out. In appearance at least, the giant salamanders resemble reconstructions of the earliest known amphibians. These were salamander-like, 7 feet (2.1 m) long, and lived 300 million years ago in the swamps that have since become coal. Another famous giant salamander lived in Europe about 12 million years ago. It was 3½ feet (1 m) long and a close relative of the three surviving giant salamanders. Some fossils of the genus *Andrias* are 7½ feet (2.3 m) long.

Adult hellbenders are remarkable for their different ways of breathing, including through internal gills, via the skin and by taking air into the lungs.

HERMIT CRAB

walking, but the next two pairs are small and are used to grip the shell. The last pair of limbs on the abdomen, which in a lobster form part of the tail fan, are sickle-shaped and used for holding on to the central column of the shell. The hermit crab has swimmerets (small, unspecialized appendages under the abdomen) only on the left side.

Shore-dwelling young

The common hermit crab, *Pagurus bernhardus,* is found in European and North American coastal waters. Normally only the young are found on the shore, their red and yellow front ends projecting from shells such as winkles, top shells and dog whelks. They are nimble in spite of their burdens and are well protected from the pounding of waves, and from drying up when the tide is out. Nevertheless, the young prefer to live in pools in most situations, and are often found under seaweed and large stones. The older ones of this species reach a length of 4 inches (10 cm), although some individuals grow to as much as 6 inches (15 cm). They live in deeper water and occupy the larger shells of common and hard whelks.

Creative shelter

Semiterrestrial hermit crabs of the genus *Coenobita* live on tropical coasts. They usually occupy ordinary snail shells, but Indonesian coenobites have been seen wearing such odd substitutes as joints of bamboo, coconut shells and even a broken oil lamp chimney. *C. diogenes,* of Bermuda, lives in shells that are in fact fossil or subfossil since they belonged to a snail, *Livona pica,* now extinct in Bermuda. The shell of another hermit crab, *Pylopagurus,* becomes encrusted with a bryozoan (moss animal). It is thought that the actual shell is dissolved, leaving only the moss animal's chalky skeleton. It is this skeleton that cloaks the crab and grows with it.

The hermit crab, *Pylocheles,* found in deep water in the Indian Ocean, lives in pieces of bamboo. Its abdomen is straight. *Xylopargus* of the Caribbean is also adapted to its living quarters. It is found at depths of 600–1,200 feet (183–366 m) in hollow cylinders of wood. The rear end of its body is shaped to make a kind of stopper.

Hermit crabs (Pagurus megistas, above) are well adapted to their shell-dwelling habit, their abdomens often being soft and twisted to fit the shells in which they live.

HERMIT CRABS LIVE IN abandoned sea snail shells. In all species the form of their bodies is modified accordingly. The banana-shaped abdomen, protected in its "hermitage," is soft and curves to the right to fit the inside of the snail shell. The front end of the hermit crab's body has the hard covering typical of crabs and lobsters, and the right claw is larger than the left in most cases. There are a few exceptions where the claws are equal, or the left is larger than the right. The crab uses its larger claw to close the entrance of the shell. It has two pairs of legs behind the claws, which are used in

COMMON HERMIT CRAB

PHYLUM	**Arthropoda**
CLASS	**Crustacea**
ORDER	**Decapoda**
FAMILY	**Paguridae**
GENUS AND SPECIES	***Pagurus bernhardus***

LENGTH
Usually less than 4 in. (10 cm); occasionally up to 6 in. (15 cm)

DISTINCTIVE FEATURES
Crab that lives in empty gastropod (sea snail) shells; soft, banana-shaped abdomen, twisted to right to fit shell; right claw usually much longer than left; 2 pairs of walking legs; 2 pairs of smaller legs for gripping shell. Adult: often found in whelk shells. Juvenile: in variety of small shells.

DIET
Tiny scraps of dead animals and plants

BREEDING
Age at first breeding: 1 year or more; breeding season: all year; number of eggs: 10,000 to 15,000; larva undergoes 4 molts; juvenile then emerges to seek suitable empty shell for a home

LIFE SPAN
Up to 4 years

HABITAT
Adult: coastal waters down to about 260 ft. (80 m), occasionally to 1,640 ft. (500 m); most seabed types except soft mud. Young: between tidemarks; especially in pools, under seaweed or large stones.

DISTRIBUTION
North Atlantic coasts including North America, western Baltic and Mediterranean

STATUS
Common

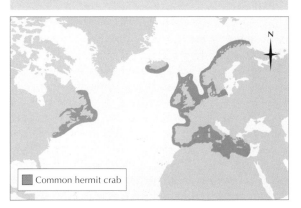

Common hermit crab

Some marine hermit crabs have less mobile homes. They live in holes in coral or sponge. This is a habit to some extent shared by lobsters and perhaps indicates the origin of the hermit crab's way of life. The robber or coconut crab, *Birgus latro*, of the South Sea Islands, is a land-living hermit crab several pounds in weight and 6 inches (15 cm) across. The adult of this species has lost the shell-dwelling habit, and although the abdomen is still twisted it has a hard covering and is kept tucked under the thorax. The robber crab makes burrows at the bases of coconut trees and lines them with coconut husks.

Hermit scavengers

Hermit crabs are mainly omnivorous scavengers, tearing up food with their smaller left claws and transferring it to their mouths. The common hermit crab feeds on tiny animals and plants, tossed with sand and debris between its mouthparts using its left claw. Some other hermit crabs can filter particles from the water with bristles on the antennae. Every so often they wipe the antennae across the mouth to take the food collected. The land-living coenobites, meanwhile,

Here a young common hermit crab can be seen moving into the empty shell of an edible periwinkle.

Here three anemones have attached themselves to the shell of a large hermit crab. The anemones benefit from being able to feed on the crab's leftover food.

The semiterrestrial hermit crabs, *Coenobita,* and the robber crab, *Birgus,* must still visit the sea to hatch their eggs because their larvae are marine. Although the adult robber crab does not carry a shell, its young will do so as they come ashore, only shedding their shells later on.

Moving house

Periodically, the growing hermit crab sheds its external skeleton. A split appears on the abdomen and the crab wriggles out of its old skin. As the hermit crab outgrows its home, this must be replaced with a larger one. The crab examines the new shell with its claws for several minutes, and then, if it seems good enough, hurriedly transfers its abdomen from the old shell to the new one. Sometimes one hermit crab may try to drive another crab from its shell.

Fascinating partnerships

Like any hard object lying on the seabed, the shell of a hermit crab tends to become encrusted with weeds, sponges, barnacles and hydroids (marine invertebrates including simple jellyfish and hydras). Certain sea anemones also associate with hermit crabs and form close partnerships with them. Known as symbiotic relationships, these are of mutual benefit to the animals. Large specimens of the common hermit crab, for example, often carry one species of anemone, *Calliactis parasitica,* on their shells. As the hermit feeds, the anemone sweeps the seabed with its outstretched tentacles and gathers fragments left by the crab. The hermit crab may sometimes benefit from bits of food caught by the anemone.

Another hermit crab, Prideaux's hermit crab, *Pagurus prideauxi,* which is light reddish brown in color and 2 inches (5 cm) long, regularly carries the anemone, *Adamsia palliata.* This species, unlike *Calliactis,* is found only on hermit crab shells. The basal disc of the anemone wraps tightly around the shell, completely enclosing it. As the crustacean grows, so does the anemone, adding to the effective capacity of the shell. As a result the shell does not have to be replaced. In this case, the mouth of the anemone lies just behind that of the hermit crab. Anemones are armed with stinging cells, and these help protect the hermit crab, discouraging, for instance, the attacks of predatory octopuses and squid.

Paguropsis typica goes a stage farther than Prideaux's hermit crab by carrying the anemone, *Anemonia mammilifera,* without the need for a snail shell as well. Another species of hermit crab, *Parapagurus pilosimanus,* has large eyes in spite of the fact that it lives in water too deep for light to penetrate. It has been suggested that *Parapagurus* finds its way about by light emitted from the phosphorescent anemone that cloaks it.

often climb bushes for plant food and may even attack young birds. The robber crab feeds on coconuts that have cracked open falling from trees, along with carrion, fruits and sago pith. It, too, is a climber and can scale the trunks of sago palms and other trees.

Breeding and growth

The common hermit crab breeds through much of the year and females with between 10,000 to 15,000 dark violet eggs attached to the swimmerets on their abdomen are to be found at most times. Such crabs come partially out of their shell from time to time and fan their swimmerets to aerate the eggs.

As the larvae hatch, molting at the same time to become zoea larvae, the mother sits partly out of her shell and gently wipes the swimmerets with a brush of bristles on her small fourth pair of legs. The tiny, shrimplike zoea larvae shed their skins four times, growing each time. At the fourth molt the young hermit crab is ready to seek its first shell home. This stage lasts 4 to 5 days. Sexual maturity is not reached for a year or more. The sexes differ externally only in the form of the swimmerets, which have differing functions. However, in many species the male hermit crab is larger than the female.

HERON

THE HERON FAMILY ARDEIDAE comprises the bitterns, night herons, tiger herons, egrets and boat-billed heron, *Cochlearius cochlearius*, as well as the many species of herons, which are discussed here. The bitterns belong to the subfamily Botaurinae, and all the other groups are placed in the subfamily Ardeinae. Bitterns' necks are shorter and thicker than those of the Ardeinae species and their bodies are generally rather smaller.

One of the most widespread species of heron is the gray heron, *Ardea cinerea*, which ranges over much of Europe and Asia as far south as Indonesia. It is also found in parts of Africa and on Madagascar. Closely related to it are the great blue heron, *A. herodias*, of North America, the Caribbean and the Galapagos Islands, and the great egret or great white egret, *Egretta alba*, which ranges from the central United States south to Argentina, across parts of Europe and Asia and south through Africa and Australia. Another cosmopolitan species is the purple heron, *A. purpurea*, which occurs in southern Europe and many parts of Asia and Africa.

Many herons have characteristic long necks and legs, with long slender bills. Others, such as the North American green-backed heron or green heron, *Butorides striatus*, and the squacco heron, *Ardeola ralloides*, of southern Europe and Africa, have comparatively short necks and legs. All of the herons have a strong association with aquatic habitats such as lakes, pools, rivers, swamps, marshes, wet meadows and coasts.

Powder down

One feature typical of the herons and their relatives is powder down patches, which are also present in a few other groups of birds such as toucans, parrots and bowerbirds. Herons have three powder down patches, one on the breast and one behind each thigh. They consist of a group of downy feathers that continually crumble into fine powder. The herons use this to absorb slime collected on their feathers after feeding on fish, first rubbing it into the plumage with the bill and then scratching it out with the comblike claws on the middle toes.

Happy landings

Most herons are gregarious, nesting in colonies, often with two or more species mixing and feeding together. Different species prefer to nest at different heights but there is still some competition for nest sites. Their flight is slow and

Usually solitary, the green-backed heron is a chunky, short-legged heron that prefers streams and marshes with woodland cover.

CLASS	**Aves**
ORDER	**Ciconiiformes**
FAMILY	**Ardeidae**
GENUS AND SPECIES	***Ardea herodias***

LENGTH
Head to tail: 3–4¼ ft. (0.9–1.3 m); wingspan: 5½–6¼ ft. (1.7–1.9 m)

DISTINCTIVE FEATURES
Long, daggerlike, yellow bill; long neck; very long yellow legs. Breeding adult: mainly grayish plumage with white face, black crest, long gray plumes on neck, pink suffusion to sides of neck and rufous thighs. Nonbreeding adult: lacks plumes on crest and neck. Juvenile: black crown; no plumes.

DIET
Fish, amphibians, small mammals such as voles, shrews and rats, insects, reptiles; rarely plant matter and birds

BREEDING
Age at first breeding: 2 years; breeding season: all year (Florida), eggs laid February–May (rest of North America), eggs laid March–July (eastern Caribbean); number of eggs: 2 to 7; incubation period: 25–29 days; fledging period: about 60 days; breeding interval: 1 year

LIFE SPAN
Up to 10 years or more

HABITAT
Edges of lakes, pools and rivers, paddyfields, wet meadows, coastal mudflats

DISTRIBUTION
Breeds from central Canada south to Cuba and Mexico and on Galapagos Islands; some birds fly south as far as Colombia in winter

STATUS
Locally common

An adult gray heron in full breeding plumage catches a fish. Herons catch prey with a sudden thrust of the bill, opening it at the last moment to seize the victim. They do not spear or stab prey.

sedate with deliberate, ponderous wingbeats. The head is drawn back and the legs trail behind. It is a most impressive sight to see a heron land in the uppermost branches of a tree, a sight that at first seems impossible. The approaching heron glides with its legs lowered and then "backwaters" with its wings until almost hovering, and gently lowers itself onto a branch. When it has found a good foothold and is well balanced, the heron folds its wings and stands still, becoming difficult to see unless silhouetted against the sky.

Much time is spent on the ground roosting with one leg raised and the head sunk into the shoulders or waiting alertly for prey on a bank or in shallow water. At other times herons will hunt actively, walking with long deliberate steps and neck stretched upward with only a trace of a kink. Occasionally a heron will swim across deeper water.

Fish is the main course

Herons eat mainly fish but their diet is varied. Water voles and frogs are probably the next most important foods of the gray heron, and other herons probably feed on similar aquatic animals. Crabs, prawns, shrimps, beetles, wasps, worms and snails are among the invertebrates taken and, perhaps surprisingly, quite a number of birds and mammals are attacked. These include thrushes, gallinules, rails, crakes, young pheasants, ducklings, rats, shrews, voles, moles and young rabbits. There are even a few records of bats being taken.

Prey is caught with a sudden thrust of the bill but is seized in the bill rather than stabbed in the manner of darters (genus *Anhinga*). The short-legged herons, such as the green-backed

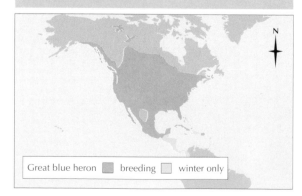

Great blue heron ◼ breeding ☐ winter only

often dive for their food from a floating log or some other perch. Indigestible remains of the prey are disgorged and can be found on the ground beneath roosts or nests. Examination of remains gives a false idea of the heron's diet because fish bones are easily digested and are not found as commonly as remains of other prey.

Blushing suitors

Several kinds of heron and egret change color in the breeding season, not by donning a fresh plumage but by changing colors of the "bare parts" such as the legs, bill, eyes and naked parts of the head. The iris and legs of the green heron change from yellow to orange. The base of the bill of the squacco heron changes from yellowish green to blue. The bill and legs of the gray heron and some other species sometimes flush red during moments of excitement. This is apparently caused by the same mechanism as blushing in humans; the blood vessels just under the skin swell and extra blood flows through them.

During courtship there are several quite elaborate displays, either at the nest or in the air, involving movements of the neck and wings and loud snapping of the bill. The nest is usually built in a high tree but is sometimes on the ground, on cliff ledges or in reed beds. A hundred or more pairs may nest together, and several nests may be built in one tree.

The nest is a platform of twigs or reeds, collected by the male and placed in position by the female. It is often used year after year, being repaired and extended each spring until it is completely blown down by the wind. In the British Isles the gray heron lays eggs in February or March and by May or June the nests are deserted. Both parents incubate the two to seven, usually three to five, eggs for about 25 days, and the young herons fly when nearly 2 months old. The great blue heron lays its eggs from February to May in most of North America, or year-round in Florida. Its clutch sizes and incubation times are the same as those of the gray heron.

When they hatch, the young are covered in down and stay in the nest until their feathers have grown. They then clamber out onto nearby branches and wait, motionless, for their parents to bring food. As the parent lands, a chick grabs its bill crosswise with its own, and food is taken from the parent's bill into the chick's.

Few enemies

Adult herons are not likely to be bothered by many predators, but their nests are sometimes robbed by crows. Herons may be harried by these birds, which either dive-bomb the herons as they stand on the ground or a perch or fly after them tweaking their tails.

Hunting strategies

Animals can be conveniently divided into generalized or specialized feeders. Some of the specialized feeders may take only one kind of food, whereas generalized feeders will eat a variety of foods, and this allows them to survive changing conditions that might wipe out a specialized feeder. Herons are not limited to eating fish, and neither are they limited to lying in wait or stalking their prey. A line of herons has been seen to beat across a field in search of mice, and they have been seen dropping mussels from a height to crack open their shells on the ground. More remarkable is the habit of the black heron, *Egretta ardesiaca*, which stands in shallow water with wings spread. One theory is that the wings provide a shade so the heron can peer into the water without being blinded by glare from the sun; another is that fish are attracted to seek shelter under the shade, only to be picked off.

A family of great blue herons in the United States. Heron chicks are covered with down at first and stay in the nest until their feathers have grown.

HERO SHREW

Hero shrews were named by an African tribe, the Mangbetu, for their great strength. It is said that a single shrew can take the weight of a person standing on its back.

SOMETIMES CALLED THE armored shrew, the hero shrew has the most remarkable vertebral column in the whole animal kingdom. The puzzle is to know what value this can have for the animal itself, something we may learn when the species has been better studied. At present there is little information on this shrew's behavior, breeding or way of life.

Hero shrews are quite large, 4¾–6 inches (12–15 cm) long, of which up to 3¾ inches (9.5 cm) is naked tail. On the rest of the body the fur is long for a shrew, coarse and thick. Its color is grayish with a slight tinge of buff. Otherwise the hero shrew looks like any ordinary shrew, with a tapering, mobile snout, small eyes and ears half buried in its fur. However, it can be distinguished by its "trotting" rather than the usual crawling gait and by its arched posture.

The single species of hero shrew, *Scutisorex somereni*, lives in rain forests in Central Africa in the Democratic Republic of the Congo (formerly Zaire), Rwanda, Burundi and Uganda.

At home under dead leaves

Hero shrews are shy and seldom seen in their homes among the dense leaf litter covering the floor of the rain forest. Their movements lack the restless energy seen in most shrews, being rather more deliberate. They occasionally show themselves when crossing a road or path, or when they come out of the leaf litter to run for the next patch of dense shadow. These shrews may be seen at various times during the day but have also been trapped at night. It is thought that they have alternating spells of activity and rest during each day, like other small mammals.

Drinking the dew

Because they have been so little studied, the diet of hero shrews has been surmised from examining the stomach contents of a few trapped individuals. These contained portions of various small invertebrates, including caterpillars, earthworms, mollusks, grasshoppers and very small frogs. The diet of hero shrews is thus probably much the same as that of other shrews. Although hero shrews can swim and climb, most food is taken from the forest floor. Prey is probably caught and immobilized by nonlethal bites. Hero shrews also lick dew from the tufts of grass and the margins of leaves. One was seen to press down the grass with its forefeet until it could easily reach the drops of water at the tips of the blades.

Very little is known of their breeding habits except that they probably breed year round and there are one to three young in a litter.

HERO SHREW

CLASS	**Mammalia**
ORDER	**Insectivora**
FAMILY	**Soricidae**
GENUS AND SPECIES	*Scutisorex somereni*

ALTERNATIVE NAME
Armored shrew

WEIGHT
1–3 oz. (30–90 g)

LENGTH
Head and body: 4¾–6 in. (12–15 cm); tail: 2¾–3¾ in. (7–9.5 cm)

DISTINCTIVE FEATURES
Very strong vertebrae (backbone) due to dorsal and ventral spines give shrew an arched posture. Tapering, mobile snout; small eyes and ears; fur is long for a shrew, coarse and thick; grayish in color with tinge of buff; naked tail.

DIET
Mainly invertebrates including caterpillars, earthworms, mollusks and grasshoppers; occasionally small frogs

BREEDING
Poorly known. Breeding season: probably all year; number of young: 1 to 3.

LIFE SPAN
Not known

HABITAT
Tropical rain forest, only at low and medium altitudes

DISTRIBUTION
Restricted range in Central Africa: northeastern Democratic Republic of the Congo (formerly Zaire), Rwanda, Burundi and southern Uganda

STATUS
Not known, but probably not threatened

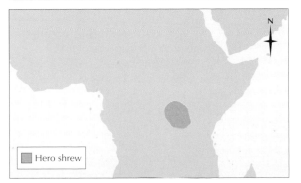

N

Hero shrew

Puzzle of the backbone

By far the most fascinating aspect of the hero shrew is its backbone. Apart from a roughening of its surface, the skull is like that of any other shrew, as are the limb bones, the tail bones and the ribs, and the shoulder and hip girdles. The neck bones are also normal, but just behind the shoulder the backbone rises in an arch as far as the hip girdle. It is several times thicker than the backbone of other shrews of similar size. This is because it has both dorsal and ventral spines. The vertebrae of the lower back are much broadened and their surfaces are ornamented with spines, ridges and bosses. As a result of these, the neighboring vertebrae interlock to give the shrew a remarkably strong arched posture.

One explanation for this enormously strengthened arch is that it might be used in turning over heavy stones as the shrews search for food and prey items underneath. They have been seen to turn over large pieces of bark and pebbles, but there is little to show that the hero shrew is any more gifted in this than other shrews or any other small mammals, such as moles. Many small mammals seem to be able to turn over heavy objects out of all proportion to their weight, and they do so without such a massive backbone as the hero shrew.

Can take the weight of a man

The name hero shrew derives from its use as a talisman or charm by the Mangbetu tribe, whose name for the shrew is a direct translation. It was also this tribe, of the Democratic Republic of the Congo (formerly Zaire), that first called the attention of scientists to the enormous strength of the shrew's backbone. Whenever they had the chance these people would show to a fascinated crowd the shrew's extraordinary resistance to pressure. A man weighing 160 pounds (73 kg) or more would stand barefoot on the shrew, balancing himself on one foot but taking care that his weight was not on the animal's head. He would hold this position for several minutes. Such a performance would squash the life out of any other shrew in a matter of seconds. Yet once the man's foot was removed, the hero shrew merely shook itself and started to walk away, to the cheers and shouts of the audience.

Because of the hero shrew's great strength, the Mangbetu are also convinced the animal has other magical properties. They believe that its charred body or even just its heart, when prepared by their ritual healers and worn as a talisman or taken as a medicine, will endow a person with heroic qualities. They also believe that such charms render them immune to serious injury from spears or arrows or from attacks by animals, including elephants.

HERRING

The Atlantic herring is the world's most important food fish. It is sold fresh, salted, smoked and canned.

No SINGLE FISH HAS HAD more influence on the course of human history or existed in greater abundance than the Atlantic herring. In the 1960s it was estimated that 3 billion were caught in the Atlantic and adjacent seas each year. The Atlantic herring belongs to the Clupeidae, a very large family containing about 215 species, including the Pacific herring (*Clupea pallasii*), American shad (*Alosa sapidissima*) and menhaden or pogy (*Brevoortia tyrannus*).

The Atlantic herring is what might well be called the typical fish, with its torpedo-shaped (fusiform) body, forked tail fin, single dorsal and anal fins, pectoral fins on the breast and pelvic fins in the pelvic region, a very prototype of fish form. Up to 16 inches (41 cm) long, it is gray green to bluish on the back and silvery on the lower flanks and belly. Its scales have only a delicate layer of skin and are readily rubbed off.

Shoals at the surface

Herrings are pelagic fish, that is, they spend much of their lives near the surface. They live in schools, each fish spaced evenly in the school with room to swim but not to turn, and they feed by taking into the mouth water that passes across the gills. The plankton in it is strained off by a fine meshwork formed by the gill rakers and swallowed. Herring have small, feeble teeth.

Moves with the currents

It was once thought that these vast shoals of herrings migrated from north to south, and the fishing fleets put out from successive ports to catch them. Now we have a different picture. Each year the Gulf Stream moves northeast across the Atlantic, reaching successively in summer the coasts of France, the British Isles, the Low Countries, Scandinavia and Iceland. Areas under the influence of the Gulf Stream are too warm for herring, which flourish in water temperatures of 43–59° F (6–15° C). However, when in summer and autumn these warm waters withdraw, shoals of herring appear, first to the north of Scotland, then in successive areas in the North Sea and finally off the coast of northwestern France.

Different races

We now know that the Atlantic herring exists in a number of races (subspecies) distinguished by the number of vertebrae, speed of growth and age of sexual maturity. Also, Icelandic, Norwegian, North Sea and English Channel herring can be recognized, and each of these includes forms that spawn at different times of the year. There are winter-spawning herrings, shedding their eggs close inshore, and summer spawners, laying in deeper waters. The pattern is complicated further because the different races migrate to a varying extent to spawning grounds or to feeding grounds. The race that spawns at the entrance to the Baltic Sea remains within that area. The Norwegian race may move from southwestern Norway northward into the Arctic, into the Barents Sea, and back again.

Mass spawning

The schools of herrings are most compact when they are made up of young fish and when adults are coming together for spawning. The act of spawning is random, the females shedding their eggs, the males shedding their milt (sperm-containing fluid) to fertilize the eggs. The adults then move on and pay no more attention to the fertilized eggs. Spawning appears to be accompanied by some excited swimming about, but there seems to be no courtship.

The eggs, 0.9–1.5 millimeters in diameter, are laid in sticky clumps that are heavier than seawater and sink to the bottom, coming to rest on the shingled seabed. Only 30,000 to 600,000 are laid by each female, a small number compared with the millions laid by some other marine fish. This is a sure sign that the eggs are relatively immune from attack, lying on the shingle beds, as

ATLANTIC HERRING

CLASS	**Osteichthyes**
ORDER	**Clupeiformes**
FAMILY	**Clupeidae**
GENUS AND SPECIES	***Clupea harengus***

LENGTH
Up to 16 in. (41 cm)

DISTINCTIVE FEATURES
Typical fish with torpedo-shaped body, single dorsal and anal fins and forked tail fin; bright greenish or bluish above; silvery lower flanks and belly with golden tints

DIET
Plankton, including crustaceans such as copepods and shrimps; also small fish

BREEDING
Age at first breeding: 3–9 years; breeding season: varies according to region but generally in warmer months; number of eggs: 30,000 to 600,000; hatching period: 8–9 days at 52–58° F (11–14° C); breeding interval: usually 1 year but some subspecies die after first spawning

LIFE SPAN
Up to 25 years

HABITAT
Mainly near surface of coastal waters

DISTRIBUTION
North Atlantic, Baltic Sea and White Sea (Arctic waters to north of European Russia)

STATUS
Common; local declines due to overfishing

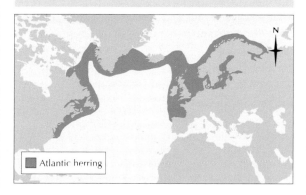

Atlantic herring

compared with floating eggs. The eggs hatch in 8–9 days in water temperatures of 52–58° F (11–14° C) but take 47 days at 32° F (0° C).

The larvae, 5–9 millimeters long when hatched, are transparent and still carry the remains of the yolk sac. They have no mouth or gills and only a single fin down the middle of the back and around the rear end. Development is rapid, however, and in a month the young fish may be ⅖ inches (1 cm) long and looks almost like its parents. The growth rate then begins to slow down, the young fish being at most 2 inches (5 cm) long, usually much less, by the end of the year. Maturity is reached in 3–5 years in the coastal waters of the North and Baltic Seas and in 3–9 years in the oceanic waters of the North Atlantic and White Sea.

Herrings founded cities...

The fishing grounds of the Northern Hemisphere, as well as supplying nations with food, have greatly influenced human history. The Atlantic and Pacific herrings are outstanding examples. It has been suggested that wherever the shoals of Atlantic herrings came in toward the coast of Norway there sprang up a village. The same seems to have been true for Scotland and Newfoundland, and for Alaska, Japan and Siberia with respect to the Pacific herring. Some of the villages have since become towns. Charlemagne, in 809 C.E., founded Hamburg as a herring port. In what is now France the Normans established Dunkirk, Etaples, Dieppe and Fécamp for the same purpose.

...and caused wars

Along the coast of North Prussia (Germany), and extending north to Norway and south to Belgium, were many free cities and small states

Herrings gather in vast shoals and are found mainly in surface waters. A shoal of one of the Pacific species is pictured off the coast of Baja California.

Atlantic herring feed on a range of plankton, especially crustaceans. In turn they provide food for larger fish such as cod, salmon and tuna, for dolphins, seals and other mammals, and for seabirds such as boobies and gannets.

carrying on general trade that were compelled to supply armed escorts for their merchandise, especially against pirates. In the 13th century they banded together to sail their great merchant fleets in convoy under protection. This cooperative group became known as the Hanseatic League, with Lübeck as headquarters. Their ships carried herrings from the Baltic ports and brought back wool, timber, wine and other merchandise. The herrings were fished by Danes off the southern coast of Sweden, but the curing and exporting were the concern of merchants in the northern German towns. The League monopolized almost the whole of the export trade of Europe and for two centuries was a dominating influence in northern Europe.

Then suddenly the stocks of herring in the Baltic disappeared, the result, it was supposed, of some natural catastrophe, now believed to have been a lowering of the water temperature. In any event, the stocks never recovered. But about this same time the Dutch had begun to export salted herrings fished in English waters. This new fishery prospered and in 1610 Sir Walter Raleigh estimated that the Dutch employed 3,000 ships and 50,000 people in their herring industry. The Dutch fishing led to friction with England, which wanted to extract a tribute for herrings taken in English waters, and the friction led also to the founding of the Royal Navy in Stuart times and to the 1652–1654 war in which England wrested sea power from Holland.

Continuing disputes

In the 19th century friction arose between the fishers of New England and those of Newfoundland over the fishing on the Grand Banks, which the Newfoundlanders regarded as their natural right. In 1877, under the Halifax Commission Treaty, the United States paid Great Britain 5.5 million dollars for their fishers to be able to fish for herring within the 3-mile (5-km) limit off the Gulf of St. Lawrence and Newfoundland. Nevertheless, one Sunday morning in Fortune Bay, the Newfoundland fishers cut the nets of two of the New England schooners, so the entire catch was lost; another New England schooner saved its catch only by threatening to shoot, and the rest of the fleet sailed for home. The incident, known as the Fortune Bay Riot, is another example of the constantly recurring friction over fishing.

Even the Russo-Japanese war of 1902 was inspired by a Japanese claim to the Pacific herrings off the Russian territory of Sakhalin Island. More recently, British herring stocks have been disputed by European fishing fleets.

HERRING GULL

THE HERRING GULL IS A familiar bird found along the shores of three continents. In Europe and North America it can be seen perching along the seafront, fighting noisily over refuse or following fishing boats in the hope of an easy meal. It is quickly distinguished from most other gulls by its size and by its gray back and wings. The wingtips are black with spots of white, the legs are flesh-colored and the bill is yellow with a red spot on the tip. Juvenile herring gulls are mottled brown all over; as they grow older the mottling gradually diminishes until, at 4 years, they attain their first adult plumage.

In North America herring gulls breed to the north of a line running approximately from the Great Lakes to the borders of Alaska. Iceland was colonized by herring gulls in 1927 and Spitsbergen in 1950. This spread is probably due to the slow warming that is taking place in the North Atlantic. Herring gulls live in much of northwestern Europe, around the coasts of the British Isles, the Faeroe Islands and eastern Iceland and on the Continent from western France to northern Finland, including the Baltic coast. They also breed in eastern Siberia.

More rubbish means more gulls
The herring gull has always been common but especially in the last 40 years its numbers have increased greatly. This is probably because of the increased amount of food in the form of edible refuse at rubbish dumps and offal at fishing ports. Some breeding colonies now number 20,000 pairs. Herring gulls have also begun to breed far inland, nesting by lakes or even on buildings in town centers.

Scavengers and ruthless pirates
Although most gulls are basically fish-eaters, many of them have made use of other sources of food. Herring gulls have become scavengers and hunters as well as fishers. They feed on any edible garbage, a very wide variety of animals and some seeds and roots. Herring gulls are a scourge of many other birds nesting in the open around coasts because they plunder nests for eggs and chicks. They also take adult shearwaters and other gulls, as well as some land animals such as shrews, rabbits, frogs and even snakes. However, the gulls probably concentrate their attacks on weakened or wounded individuals.

The shore is a favorite feeding ground for herring gulls. They search among rocks and in pools for shellfish and crustaceans. Shellfish such as mussels, which can protect themselves by

closing up, unlike rock-hugging limpets, are carried into the air and dropped so the shell is broken. This seems to be proof of considerable intelligence but the gulls do not choose where to drop them. They can be seen flying up 20 feet (6 m) or so, repeatedly dropping shells onto soft sand. It is only when the shells drop onto rocks, quite by chance, that the method is successful.

Dive-bomber camouflage
When fishing at sea, herring gulls, like many of their relatives, catch their prey by plunge-diving, dropping into the water from a few feet above with wings half folded. The gulls very rarely submerge completely, usually immersing only head and neck or part of the body. This means that they must catch the fish they are aiming at the first time, or else the prey will escape. Experiments at Oxford University, England, have shown that the essentially white plumage of a herring gull is an advantage in this respect. By rolling models down a slope toward an aquarium it was shown that the fish in the aquarium reacted more quickly to a dark model. They were able to flee several inches farther before a dark model struck the water than they could if it was white. Observations in the North Sea suggest that this is important in practice. Adult herring gulls are found mainly out to sea, where plunge-diving is required, while juveniles, with their dark plumage, stay near the shore to feed on refuse and shore creatures.

Extremely efficient scavengers of the seashore, herring gulls have been spreading far inland to take advantage of the rich pickings to be had at garbage dumps.

Herring gulls acquire their first white and gray adult plumage when 4 years old. The red spot on the bills of the adult gulls is pecked at by their chicks whenever they are hungry.

Herring gulls of all ages also hunt for food on farmland, particularly outside the breeding season. They hunt earthworms, beetles and other small prey as well as eating spilt grain. Garbage dumps are a rich source of food, and some herring gulls may feed nowhere else. In Britain during the 1980s and 1990s, urban populations of herring gulls were thriving, whereas there was a decline in some coastal populations.

Nest in colonies

Herring gulls nest in colonies on cliffs or small islands or among sand dunes. Colonies can also be found inland, by lakes and reservoirs, in bogs and marshes or on buildings. Each pair has a small territory surrounding their nest of grass, seaweed or other plants. Each gull returns regularly to the same territory with the same mate. Herring gulls spend the winter away from the colony; then in early spring they return to claim their territories, and a certain amount of fighting takes place until they have settled down. Courtship then gets under way, either to renew old ties or to find a new mate if the old one has died.

After the nest is built, usually three brown eggs with blackish blotches are laid. Clutches of two or four eggs are less common. Both sexes incubate them, taking turns of a few hours each for about 28–30 days. In common with many birds that nest on the ground, herring gulls will retrieve their eggs if they are knocked out of the nest. This is quite easy to demonstrate by putting one of the eggs just outside the nest. The adult

HERRING GULL

CLASS	**Aves**
ORDER	**Charadriiformes**
FAMILY	**Laridae**
GENUS AND SPECIES	***Larus argentatus***

LENGTH
Head to tail: 21½–25 in. (55–64 cm); wingspan: 55–59 in. (1.4–1.5 m)

DISTINCTIVE FEATURES
Stout, slightly hooked bill; fairly long wings; webbed feet. Adult: mid-gray upperparts; white head and underparts; black wingtips; pink legs; yellow bill with red spot near tip. Juvenile (1–4 years): mottled brown all over, becoming whiter with age; brownish gray bill with black spot near tip.

DIET
Marine animals such as fish, squid, mollusks, crustaceans, seabird eggs and nestlings; plant matter such as seeds and roots; land animals such as earthworms, shrews, frogs and young rabbits; carrion and garbage

BREEDING
Age at first breeding: 4–5 years; breeding season: eggs laid mid-April to mid-May; number of eggs: usually 3; incubation period: 28–30 days; fledging period: 35–40 days; breeding interval: 1 year

LIFE SPAN
Up to 32 years, usually much less

HABITAT
Mainly coasts and estuaries; also inland: farmland, reservoirs, gravel pits, lakes, garbage dumps, towns and cities

DISTRIBUTION
Breeds in Canada, northeastern U.S., Iceland, northwestern Europe and Siberia; some populations move south in winter

STATUS
Common or very common in much of range

Herring gull ▇ all year or summer only ▢ winter only

gull returns and settles down on the nest, but it realizes something is wrong. It stands up again and looks at the eggs and shuffles them about. Eventually the gull leans forward, hooks its bill over the stray egg and draws it into the nest, and all is well.

Hitting the spot

The chicks are brooded in the nest at intervals while they are young, but from a very early age they can run about the nesting territory. The parents bring food back to them. At first the adults regurgitate it and hold small pieces in the tip of the bill for the chicks to peck at. Later, food is dropped to the ground and the chicks pick off pieces for themselves. Herring gulls, as well as other gulls, terns and jaegers, are stimulated to feed their chicks by the latter pecking at the parents' bills. In the herring gull, the chicks aim at the red patch on the bill, and it has been shown that in all these birds the chicks are specially sensitive to red, so the red spot serves as a "target" for the chick to peck at.

Parent gulls defend their chicks vigorously, flying at intruders to scare them away and giving alarm calls that alert the chicks and send them scurrying to shelter. The chicks start to fly when 6 weeks old but take a few days to learn efficient takeoff and landing. A short time later they leave the colony, although they will stay with their parents for a little time.

Around the world

The lesser black-backed gull, *Larus fuscus*, lives around the coasts of Europe and across Asia. It is very similar to the herring gull except that its back and wings are dark gray to black and the legs are yellow. Also, the herring gull has a ring of yellow flesh around the eye, while the lesser black-backed gull has a vermilion eye ring. These physical differences are sufficient to class the two as separate species, but herring gulls and lesser black-backed gulls occasionally interbreed.

In fact, the herring and lesser black-backed gulls together make up what is known as a ring species. Around the Northern Hemisphere there are numerous races, or subspecies, of both forms. The races of each form are sufficiently like neighboring races to interbreed freely with them, but the two streams have diverged and where they meet and overlap in northwestern Europe interbreeding between the herring gull and lesser black-backed gull is very rare.

Like other closely related species, the herring and lesser black-backed gulls also have isolating mechanisms. Despite the fact that they nest together in mixed colonies, they are kept in practical isolation by their behavior. For example, the courtship calls and displays of the males of each species are somewhat different, so it is likely that the female gulls are able to select the right mate, aided also by visible differences in wing and eye ring color.

Herring gulls often follow ships out to sea and gather at ports and fish-processing plants to scavenge scraps.

HIDDEN-NECKED TURTLE

WHEN A TURTLE OR TERRAPIN pulls its head into the shelter of its shell, its neck is usually bent vertically. The hidden-necked turtles, often called the side-necked or snake-necked turtles, bend their necks horizontally. There are approximately 60 species of these turtles, although the exact number is in dispute. The neck is very long in some species, but very short in others, for example the North Australian snapping turtle. Hidden-necked turtles are found in South America, except in the extreme south; sub-Saharan Africa, including Madagascar; Australia, except in the arid middle and in Tasmania; and New Guinea. They are all aquatic or semiaquatic, living in or around the margins of rivers, lakes, ponds and streams.

A number of different names are used for these animals in different parts of the world. In South America, the larger species are called river turtles, while names for the smaller species include side-necked turtles, swamp turtles and toad-headed turtles. African forms are often called mud turtles, while in Australia and New Guinea they are referred to as snake-necked turtles. Because of their habits, some are also called snapping turtles. This article concentrates on two species, the Amazon river turtle, which is also found in the basin of the Orinoco River, and the African helmeted turtle, sometimes called the water tortoise or marsh terrapin.

The Amazon river turtle is the world's largest species of freshwater turtle. It is also notable for its breeding habits, which are like those of marine turtles.

Social breeding on sandbanks

The Amazon river turtle is known in the Orinoco basin of Venezuela as the *arrau*. A female can reach 3 feet (90 cm) in length, but the biggest female recorded was 3½ feet (1.1 m) long. The smaller, almost circular, male grows to an average diameter of only 1½ feet (45 cm). Most females weigh about 50 pounds (23 kg), but giant specimens of up to 130 pounds (60 kg) have been recorded.

Amazon river turtles generally behave as other freshwater turtles do, apart from their breeding. This is remarkably like that of marine turtles. In early February, as the dry season begins, the waters drop, exposing sandbanks and islands in midriver. The turtles gather in their thousands in the water around the sandbanks, where they mate. Some have traveled 100 miles (160 km) to find a suitable sandbank area. After mating, the males leave, while the females land at night to lay their eggs.

Thousands of sand-packed eggs

Each of the females digs a pit in the sand. This is usually about 2 feet (60 cm) deep and 3 feet (90 cm) across at the top. It is dug using all four legs. There is an extra depression at the bottom, dug with the hind legs only. The female then lays an average of 80 eggs, although there may be up to 150, into the depression. Each egg is slightly less than 2 inches (5 cm) in diameter and, like all turtle eggs, has a soft, parchment-like shell. After this, the female fills in the pit, disturbs the sand all around to mask the actual nest, and leaves. Night after night thousands more females arrive to lay their eggs. Six weeks later the 2-inch (5-cm) hatchlings dig their way up out of the sand to run a gauntlet of vultures, storks and ibises. Those that reach the water face other hazards: crocodiles and predatory fish.

Unloved turtle

The African helmeted turtle may grow to just over 1 foot (30 cm) in length, and is 10 inches (25 cm) wide. Its back is a mottled greenish brown. The turtle, which is found throughout Africa south of the Sahara, has earned a bad reputation in South Africa for its strong smell and its attacks on ducklings. It also steals bait from anglers' hooks. The odor comes from four glands, one under each leg, which give out a foul-smelling liquid that is said to be

HIDDEN-NECKED TURTLES

CLASS	**Reptilia**
ORDER	**Testudines**
SUBORDER	**Pleurodira**
FAMILY	**Pelomedusidae and Chelidae**

GENUS AND SPECIES **60, including Amazon river turtle, *Podocnemis expansa*; and African helmeted turtle, *Pelomedusa subrufa***

ALTERNATIVE NAMES

Amazon river turtle: arrau. African helmeted turtle: water tortoise; marsh terrapin. Various other species: river turtle, swamp turtle, side-necked turtle, toad-headed turtle (South America only); snake-necked turtle (Australia and New Zealand only); mud turtle (Africa only).

LENGTH

Amazon river turtle: up to 3½ ft. (1.1 m). African helmeted turtle: up to 13 in. (33 cm).

DISTINCTIVE FEATURES

Neck bent horizontally when withdrawn

DIET

Usually a mixture of animal and plant foods

BREEDING

Amazon river turtle. Breeding season: early February; number of eggs: usually about 80; hatching period: 42 days.

LIFE SPAN

Amazon river turtle: at least 40 years

HABITAT

Margins of rivers, lakes, swamps and ponds, including temporary pools

DISTRIBUTION

Amazon river turtle: northeastern South America. African helmeted turtle: sub-Saharan Africa and Madagascar.

STATUS

Amazon river turtle: at low risk. African helmeted turtle: common.

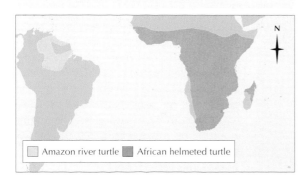

Amazon river turtle ■ African helmeted turtle

An African helmeted turtle in the Kalahari, southern Africa. These turtles are unpopular with local people due to their strong smell.

especially objectionable to horses and humans. The turtle's food is almost entirely animal, although it readily feeds on plants. It runs fast on land and swims even faster, so ducklings are highly vulnerable to being seized by a leg and dragged underwater.

Separate nests in the mud

When the female African helmeted turtle is about to lay, she comes on land, selects a site, releases a quantity of urine on the ground and puddles the mud with her feet. She repeats this several times until a stiff mud is formed, making digging easier. Then she digs a hole 4 inches (10 cm) across, at the bottom of which she excavates a smaller chamber. Into this she lays about 20 oval, soft-shelled eggs, 1½ inches (3.8 cm) long by ¾ inch (2 cm) across. The shell membrane later hardens. When the hatchlings burrow up to the surface, they face predators such as herons and other birds.

Species on the decline

Amazon river turtles have declined dramatically over the past two centuries and are now considered to be at low risk. Some species, such as the African helmeted turtle, are abundant in many parts of their range. The status of many other hidden-necked turtles is not known, although some of those with restricted distributions are known to be vulnerable. The western swamp turtle, *Pseudemydura umbrina*, of western Australia is restricted to marshes near Perth and is one of the world's most endangered turtles.

HIPPOPOTAMUS

Hippopotamus herd resting on the bank of the Mara River in Tanzania. Hippos spend much of their time in or near water and cannot survive away from it for long.

DISTANTLY RELATED TO pigs, the hippopotamus rivals the Indian rhinoceros as the third-largest living land animal, after the two species of elephants. The hippo can grow to more than 16 feet (5 m) long and can stand about 5 feet (1.5 m) at the shoulder. It weighs up to 5 tons (4,500 kg). Its enormous body is supported on short, pillarlike legs, each with four toes ending in hooflike nails, placed well apart. A hippo trail in a swamp shows as two deep ruts made by the feet, with a dip in the middle made by the belly. The hippo's eyes are raised on top of its large, flattish head. It has small ears and slitlike nostrils high up on the muzzle. Its body is hairless except for sparse bristles on the muzzle, inside the ears and on the tip of the short tail. The hippo has a thick layer of fat under its skin. Its hide is a deep purplish gray to blue-black color. Pores in the skin give out an oily pink fluid, known as "pink sweat." This lubricates the skin. Its mouth is armed with large canine tusks that average 2½ feet (75 cm) long. In males these may reach over 5 feet (1.5 m), including the long root embedded in the gums.

Once numerous in rivers throughout Africa, the hippo is now extinct north of Khartoum, Sudan, and south of the Zambezi River in southern Africa, except in a few protected areas such as Kruger National Park in South Africa. Its present distribution continues to shrink. During the dry season hippos congregate together at permanent water sources, but disperse more widely during the rains.

The pygmy hippopotamus

As recently as a million years ago there were eight or more species of hippo in Africa, all differing in size and diet. Now there are only two: the hippopotamus, *Hippopotamus amphibius*, and the pygmy hippopotamus, *Choeropsis liberiensis*. Other species are thought to have disappeared from India and Madagascar. The pygmy hippo lives in forest streams in Liberia, Sierra Leone and parts of southern Nigeria. It is 5 feet (1.5 m) long, 2⅔ feet (80 cm) at the shoulder and weighs up to 600 pounds (270 kg). Its head is smaller in proportion to its body, and it is found singly or in pairs.

The river horse

The name hippopotamus literally means river horse. The hippo spends most of its time in rivers, being unable to survive for long periods away from water. It comes on land to feed, but only at night or during rain. It can remain submerged for up to 4½ minutes and spends the day basking on a sandbar or

HIPPOPOTAMUS

CLASS	**Mammalia**
ORDER	**Artiodactyla**
FAMILY	**Hippopotamidae**
GENUS AND SPECIES	***Hippopotamus amphibius***

ALTERNATIVE NAME
Kiboko

WEIGHT
2,200–9,920 lb. (1,000–4,500 kg)

LENGTH
**Head and body: 9½–16⅖ ft. (2.9–5 m);
shoulder height: 5 ft. (1.5 m);
tail: 1⅓–1⅘ ft. (40–55 cm)**

DISTINCTIVE FEATURES
**Resembles a massive, amphibious pig;
huge head with raised eyes and nostrils on
top; long jaw with large tusks (canine teeth);
mainly hairless; short, pillarlike legs with
4 toes and hooflike nails; short tail; purplish
gray to blue black in color**

DIET
Creeping and tussock grasses

BREEDING
**Age at first breeding: mature at 3–4 years,
mating unlikely before 7 years; breeding
season: mainly February and August;
gestation period: usually about 240 days;
number of young: usually 1; breeding
interval: about 2 years**

LIFE SPAN
20–40 years

HABITAT
**Areas of deep, permanent water with
adjacent reed beds and grassland**

DISTRIBUTION
Rivers in sub-Saharan Africa

STATUS
Locally common, but vulnerable to hunting

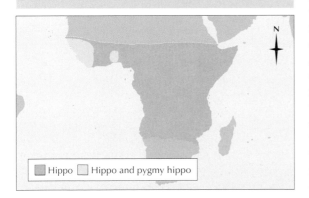

Hippo ☐ Hippo and pygmy hippo

*The pygmy hippo is
considerably smaller
and rarer than its
better-known relative.
The captive animal
shown above belongs
to a breeding program
that is aiming to save
the species.*

lazing in the water with little more than its ears,
eyes and nostrils showing above the surface. At
most its back and the upperpart of its head may
be exposed. Although there are many hippos in
national parks, they are still hunted for their
meat and because they cause damage to commercial crops. Where they are heavily hunted,
hippos keep to reed beds.

Each group, sometimes spoken of as a school,
numbers around 20 to 100 animals. The group's
territory is made up of a central crèche occupied
by females and juveniles. There are then separate
areas, known as refuges, around the perimeter,
each occupied by an adult male. The crèche is on
a sandbar in midstream or on a raised bank of the
river or lake. Special paths lead from the males'
refuges to the feeding grounds, each male
marking his own path with dung. Females also
have their own paths, but are less exclusive.

Strict social hierarchy

It was long thought that a hippopotamus school
was led by the oldest male. It is, in fact, more of a
matriarchy. Females and their young form the
only stable social unit and often dominate. There
are strict rules of behavior within the group. For
example, outside the breeding season a female
may pay a social call on a male and he may return
this, but on the female's terms. He must enter the
crèche with no sign of aggression; should one of
the females rise on her feet he must lie down.
Only when she lies down again may he rise. A
male failing to observe these rules will be driven
out by the adult females attacking him *en masse*.

Competition to mate

Female hippos also have some influence over the organization of territories. On leaving the crèche, male hippos are forced to stay beyond the ring of refuges lying on the perimeter of the crèche. From there each must win his way to an inner refuge, which entitles him to mate with one of the females, by fighting with other males. A young male might be overpersecuted by the senior males, particularly if refuges become overcrowded. Should this be the case, he can reenter the crèche for sanctuary, protected by the combined weight and position of the females.

The characteristic yawning of the male hippo is an aggressive gesture, a preliminary challenge to fight. Often competition between the males in a school can become fierce. Combats are a vigorous trial of weight and strength, the two contestants rearing up out of the water, their enormous mouths wide open, seeking to deliver slashing cuts with their long tusks. Deep gashes may be inflicted, although these normally heal quickly.

Nightly wanderings

Hippos feed mostly at night, coming on land to eat mainly creeping and tussock grasses. Their appetites are quite modest considering their size, and their broad, grass-cropping lips are usually able to gather enough food in less than 5 hours. Hippos have a slow but efficient digestion. During one night an individual may wander anywhere up to 20 miles (32 km), but it usually does not venture far from water. If alarmed, hippos have been known to run at speeds of 30 miles per hour (48 km/h).

Young in nursery school

Hippos are polyestrus, meaning that the females come into season more than once each year. Mating normally takes place in February or August. When in season the female goes out to choose her mate, and he must display deference as she enters his refuge.

Normally a single calf is born 210 to 255 days after mating. Twins are very rare. The young hippo is 3 feet (90 cm) long, 1 foot (30 cm) tall and weighs around 60 pounds (27 kg). Birth may take place in water but is normally on land, the mother first preparing a bed of trampled reeds. The calf can walk, run or swim 5 minutes after birth and is even able to suckle underwater.

Outside the crèche the organization of the school is dependent on fighting, and the females educate their young accordingly. A short while

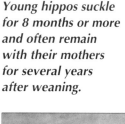

Young hippos suckle for 8 months or more and often remain with their mothers for several years after weaning.

after its birth the baby hippo is taken on land for walks, not along the paths used when going to pasture but in a random promenade. The youngster must walk level with the mother's neck, presumably so she can keep an eye on it. If the mother quickens her pace, the young hippo must do the same. If she stops, it must also stop. In water the calf must swim level with the mother's shoulder, matching her speed. Strict obedience is expected of the young hippos, and the penalty for failing to do so is severe. The mother may lash the erring youngster with her head, often rolling it over and over.

On land the lighter female is more agile than the male, so she can defend her baby without difficulty. In the water the larger male, with his longer tusks, has the advantage, so it is important for the calf to be where the mother can quickly interpose her own body to protect it from an aggressive male.

Later, when she takes it to pasture, the calf must walk at heel. If a mother has more than one youngster with her, which can happen because her offspring stay with her for several years, they walk behind her in order of precedence, the eldest bringing up the rear.

Leaving young with baby-sitters

If a female leaves the crèche for feeding or mating, she places her youngster in the care of another female, which may already have several others under her supervision. This is made easier by the fact that hippo mothers with young of a similar age tend to keep together in the crèche. The young hippos play with others of similar age, the young females together playing a form of hide-and-seek or rolling over in the water with stiff legs. The young males also play together, often indulging in mock fights.

Conservation

Hippos have virtually no predators, although a lion may occasionally spring on the back of a hippo on land, raking its hide with its claws. Young hippos, despite the protection of their mothers, are at risk from crocodiles. Humans pose the main threat to hippos of both species. Although there can be very high population numbers in small areas, hippos remain vulnerable to hunting and, in the case of the pygmy hippo, extensive deforestation presents another serious threat. Numbers of pygmy hippos, in particular, are in serious decline.

Male hippos come to blows, Lake Manyara, Tanzania. Competition for the inner refuges, and the corresponding entitlement to mate with females, is fierce.

HOATZIN

Hoatzins live in small groups made up of a breeding pair and several nonbreeding helpers. They feed mainly at dawn and dusk, and often sunbathe during the heat of the day.

APPARENTLY A LINK WITH birds that became extinct millions of years ago, the hoatzin is one of the strangest of all living birds. It is classified with the domestic chicken in the order Galliformes but it is unique in many ways, being very primitive in some respects and very advanced in others.

The adult hoatzin is the size of an American crow, *Corvus brachyrhynchos*, and looks rather like a scruffy pheasant with large wings and a long, broad tail. The small head is set on a long thin neck, rather like a peacock's, except that in place of the peacock's gleaming feathers and immaculate crest it has a long crest of bristly feathers. The upperparts are bronze-olive, streaked with white, and the underparts are white.

Restricted food supply

The hoatzin is restricted to flooded forests and woodlands along big rivers in northern South America, from Venezuela and Guyana south to Brazil. Its diet is composed almost entirely of vegetable matter. About 80 percent of its food intake consists of new growth of green leaves and buds, with over 50 species of plants recorded in the diet. Some flowers and fruits are also eaten in season. The hoatzin feeds in social groups, mainly in the early morning and near dusk. It also feeds on moonlit nights.

Many birds with such a restricted diet and habitat are threatened nowadays, but the hoatzin seems fairly secure, at least for the moment, as the flooded forests of South America are relatively undisturbed and hoatzins are said to be unfit to eat, although their eggs are taken. They have a variety of local names, including "Stinking Ana," which refers to the smell of hoatzin flesh. Apparently the unpleasantness is restricted to a musky odor, which does not affect the taste of the flesh but is enough to discourage eating it.

Strange digestive system

The hoatzin has a most peculiar digestive system, which affects its whole life. In most plant-eating birds the food is first stored in the crop (a pouchlike sac in the gullet) and then ground up in the gizzard (a muscular enlargement of the alimentary canal). The hoatzin has an imperfectly developed gizzard but the crop is much enlarged with thick walls and a horny lining, both used to grind coarse leaves. The crop is high in the chest, so a hoatzin tends to be top-heavy after a large meal. Its legs look strong but are not able to support it properly on its perch. For this reason hoatzins rest their bodies on the breastbone, which is covered by a callosity: a special pad of horny skin.

Another result of the strange digestive system is the hoatzin's weak power of flight. The huge crop takes up a large amount of room in the chest, at the expense of the flight muscles. The result is that the hoatzin can flap only feebly for short distances, perhaps 20 yards (18 m) or so. Sometimes hoatzins climb to the tops of trees to flutter across a stream or pool. They are highly sedentary, moving no more than 1¼ miles (2 km).

Helping out at the nest

Hoatzins live in small groups of two to eight birds, which stay together year-round. They often nest in colonies. These groups are composed of the breeding pair and a number of helpers. The helpers are nearly all young produced by previous nestings of the main breeding pair. Helpers perform a range of duties: territorial defense, nest-building, incubating the eggs and feeding the young. Studies have shown that nests with helpers are 45 percent more successful than those without.

The nest is a rough platform of twigs a little like a pigeon's nest, about 12 feet (3.7 m) above the water. This is by no means a safe height as the rivers are subject to severe flooding. Both sexes of the breeding pair, assisted by their helpers, build the nest. Usually two small yellowish eggs are laid, although up to four eggs are known. Mating takes place only between the

HOATZIN

CLASS	**Aves**
ORDER	**Galliformes**
FAMILY	**Opisthocomidae**
GENUS AND SPECIES	***Opisthocomus hoazin***

WEIGHT
Up to 29 oz. (820 g)

LENGTH
Head to tail: 2 ft. (60 cm)

DISTINCTIVE FEATURES
Adult: small head; naked face with bright blue skin; red eyes; bristly, fan-shaped crest; short, heavy bill; large, weak wings; long, broad tail; mainly bronze-olive with white streaks above and white below; chestnut crest, foreneck and breast. Nestling: 2 large claws on front edge of each wing.

DIET
Mainly new growth of green leaves and buds; flowers and fruits in season

BREEDING
Age at first breeding: 2–3 years; breeding season: all year with peak in rainy season; number of eggs: usually 2; incubation period: 30–31 days; fledging period: about 60 days; breeding interval: varies but usually about 1 year

LIFE SPAN
Not known

HABITAT
Wet or flooded forest; strips of woodland

DISTRIBUTION
Eastern and southern Venezuela, Guyana, Suriname and French Guiana south to Amazon Basin in Brazil

STATUS
Locally common in some regions; not threatened

Hoatzin

breeding pair, but display copulations—which take place outside the breeding season as well—involve the breeding male and any of the females, including helpers.

Acrobatic young

The young hoatzins are also truly remarkable. When they hatch they are covered with the first of two coats of down. They spend a considerable time in the nest being fed by the parents. The adult opens its mouth and the chick puts its bill in to take the food. Later the young leave the nest and crawl around the neighboring branches. Each of their wings is equipped with two claws, which the chicks use, together with their bills, as an aid to clambering. They climb about with great agility and can evade capture this way.

As a last resort the chicks leap off the branch and fall into the water. Here they continue to evade capture by repeated diving. When danger is past they climb out of the water and are cared for in low vegetation. For a nonaquatic bird to use water to such an extent is most unusual.

Primitive wing claws

The wing of a normal bird is built on the same pattern as the arms and legs of amphibians, reptiles and mammals: the pentadactyl or five-fingered limb. In the wing the fourth and fifth fingers have been lost, the second and third bear the flight feathers and the "thumb" forms the small "bastard wing." Fossils of Archaeopteryx show that this prehistoric bird had claws on the tips of each finger or digit. It had feathers but the wings were so weak that it could not have flown properly and must have lived the same kind of life as the hoatzin does today. Hoatzin chicks are even more like Archaeopteryx than the adults are. The development of their wing feathers is retarded, so the wing claws are free for hanging onto twigs and branches. As the chicks develop the claws are lost and the wing feathers grow.

The hoatzin chick has a pair of claws near the bend of each wing, which are unique in the bird world. Using a combination of these wing claws and its bill, the chick can clamber about in the branches near to its nest.

HOBBY

Hobbies can be identified by the dark streaking on their white underparts and the broad black stripe that looks like a drooping mustache.

THE HOBBY IS A SMALL FALCON with long, scythelike wings and a relatively short tail, giving it a swiftlike silhouette. When perching, the wings extend slightly beyond the tail. The hobby's upperparts are blue gray and it has a black crown that tapers into a broad black stripe like a drooping mustache. The hobby is white underneath, with conspicuous dark streaking, and it has rusty-orange thighs.

There are two species: the hobby, *Falco subbuteo*, and the African hobby, *Falco cuvieri*. The hobby is widespread across much of Europe except in northern Scandinavia, Sicily, Sardinia and other Mediterranean islands. Hobbies also breed in northwestern Africa and throughout temperate Asia to Kamchatka and the Kurils in the eastern Pacific. The African hobby is slightly smaller than the hobby, 9½–11 inches (24–28 cm) total length compared to 12–14 inches (30–36 cm).

The African hobby is found in eastern Africa from South Africa north to Ethiopia and across to Liberia.

The hobby is a summer resident in Europe, migrating in winter to Africa, where it can be distinguished from its African counterpart by a generally paler plumage. In their breeding haunts in Europe, hobbies live in open woodland, on heath and downs with scattered clumps of trees, or on farmland with hedgerows, but in their winter home they prefer open savanna country.

Taking birds from the air

Hobbies hunt prey on the wing rather than pouncing on earthbound animals as harriers or kestrels do. They dive after small birds with great speed and agility, rivaling peregrines and merlins in their aerial chases. Hobbies will often dash through flocks of swallows or starlings at breakneck speed, seizing a bird as they go, rather than chasing a selected individual. Skylarks, martins and swallows are all favored prey, but hobbies are known to take birds varying in size from tits up to the occasional pigeon or partridge.

Apart from birds, hobbies feed on many insects, which often form the bulk of their diet. Slow fliers, such as beetles, are hunted in a leisurely manner, but faster insects, such as dragonflies, are chased with great agility. Insects are eaten in the air. In Africa hobbies prey on flying termites. They also prey on a few small mammals such as shrews and mice and they occasionally take bats. On rare occasions hobbies have been seen to steal prey from kestrels.

Aerobatic antics

The aerial courtship of hobbies is even more spectacular than that of other raptors (birds of prey) such as harriers. A strident *kew-kew-kew* often calls attention to a pair of hobbies as they prepare to indulge in a display of aerobatics. The pair will soar together, circling upward until almost lost from sight and then swoop at each other, circling around or looping the loop, and occasionally gliding upside down. The culmination of these displays of aerobatic skill comes when the male brings food to the female. Male harriers also pass food to the female in flight, but the male hobby usually does so at higher speeds.

HOBBY

CLASS	**Aves**
ORDER	**Falconiformes**
FAMILY	**Falconidae**
GENUS AND SPECIES	***Falco subbuteo***

WEIGHT
**Male: 7–7⅖ oz. (200–210 g);
female: 8⅔–11½ oz. (245–325 g)**

LENGTH
**Head and body: 12–14 in. (30–36 cm);
wingspan: 2¾–3 ft. (82–90 cm)**

DISTINCTIVE FEATURES
**Long, scythelike wings; short tail; black
crown; black stripe on each side of face,
resembling mustache; blue-gray upperparts;
many dark streaks on white underparts;
rusty-orange thighs**

DIET
**Mainly small birds such as swallows, skylarks
and martins; also insects such as dragonflies,
beetles and termites; sometimes bats**

BREEDING
**Age at first breeding: not known; breeding
season: eggs laid mid-May to early June;
number of eggs: usually 3; incubation
period: 28–31 days; fledging period: 28–34
days; breeding interval: 1 year**

LIFE SPAN
Up to 11 years

HABITAT
**Open expanses of low vegetation with tall
trees or fringed by woodland; also farmland**

DISTRIBUTION
**Summer: much of Europe; North Africa;
east across Asia to eastern China. Winter:
southern Africa; India; southern China.**

STATUS
Uncommon

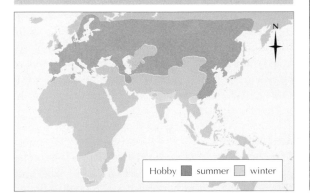

Hobby ▇ summer ▢ winter

He dives at full speed from a great height and
then soars up as he passes the female, and she
takes the prey from him.

Take over nests

Like other falcons, hobbies do not build their own
nests but take over the deserted nests of other
species, often flattening them and removing part
of the original lining. Nests of carrion crows and
other members of the crow family are most often
used, together with those of herons, sparrow
hawks, wood pigeons and tree squirrels.

The normal clutch consists of three eggs,
which are incubated mainly by the female.
Incubation takes around 28 to 31 days. The male
hobby feeds her during this period. He calls her
off the nest to give her food either at a nearby
perch or in the air.

After the chicks have hatched the female may
collect food for them in this manner, or the male
might bring food directly to the nest. The female
begins to bring food when the chicks are quite
large, unless the pair are feeding mainly on
insects. If this is the case, both hobbies hunt, bringing small amounts of food at short intervals. The
young are able to fly when about 30 days old and
are fed by the parents for a short time.

Nimble killers

Hobbies have an impressive ability to follow
every twist and turn of a prey item, such as a
dragonfly or barn swallow, in an attempt to catch
it. In falconry hobbies were once used to hunt
larks, which attempt to escape by complicated
maneuvers. Incredible aerial "dogfights" can still
be seen between hobbies and swallows or swifts,
and even dragonflies, which flash back and forth
as if on a spring.

*An adult hobby at its
nest. Like harriers, the
female incubates the
eggs, during which time
the male feeds her.
Both parents care for
the young even after
they have learned to fly.*

HOCHSTETTER'S FROG

Hochstetter's frog in the mountain forests of New Zealand. This species and its relatives are thought to be the most primitive of frogs.

HOCHSTETTER'S FROG IS chosen here as a representative of the primitive frog family Leiopelmatidae, of which three species live in New Zealand. The only other species in this family, the American tailed frog, *Ascaphus truei*, is found in northwestern America, the opposite corner of the Pacific. As its name suggests, this frog has a "tail," while the other three species do not. However, they have retained the muscles that would be required to control it. The "tail" is an extension of the cloaca (a chamber into which the urinary, intestinal and generative canals feed) and is found only in males.

In all four species, the toes are only slightly webbed. Each vertebra is concave on both faces, a condition known as amphicoelous. The vertebrae of almost all fish are also amphicoelous, whereas those of all other frogs are not. Hoch-

stetter's frog and its relatives seem therefore to be very similar to the early, fishlike ancestors of amphibians. The Leiopelmatidae not only are the most primitive of all frogs, but are also very rare. They are protected in New Zealand.

Mountaintop homes

Hochstetter's frog and its relatives are less than 2 inches (5 cm) long. They live mainly near mountaintops, where the air is moist and cold and the temperature of any water is rarely above 40° F (4° C). Their food is mainly insects, including the larvae, as well as spiders and wood lice or sow bugs. These frogs have a large tongue that is rounded or pear-shaped and almost completely fastened to the floor of the mouth, so it cannot be shot out to capture prey.

Advanced tadpoles

The eggs of Hochstetter's frog are terrestrial and can be found in winter months in the seepage from mountain streams, in tunnels in the wet clay. The tunnels are probably made by dragonfly larvae. On the floors of the tunnels the eggs, 5 millimeters in diameter when laid and strung together like beads in groups of up to 11, lie on the mud, washed by slow trickles of water. Male frogs guard the eggs before they hatch into advanced, nonfeeding tadpoles. After hatching, the tadpoles may be carried on the back of one of the parents for short periods of time.

Archey's frog, *Leiopelma archeyi*, is also found in the high New Zealand mountains. In common with the Hochstetter's frog, the male frogs guard the eggs before hatching, but this species lays its slightly smaller eggs under stones. Each clutch contains nine eggs. Both Archey's frog and Hamilton's frog, *L. hamiltoni*, have direct development, but the young hatch before their tails have been fully absorbed.

In Archey's frog, the froglets hatch after about 44 days. Each froglet breaks out of its egg by lashing with its tail, which then becomes an important breathing organ. The lungs do not develop fully for some weeks after hatching, and in the meantime the froglet breathes through the skin of the belly and tail, both of which are richly supplied with fine surface blood vessels. For the first month, the froglet lives on the remains of yolk from the egg.

Tailed frogs

The American representative of the Leiopelmatidae has been variously called the tailed frog and the bell toad. It can be found in the United States

HOCHSTETTER'S FROG

CLASS	**Amphibia**
SUPERORDER	**Salientia**
ORDER	**Anura**
FAMILY	**Leiopelmatidae**
GENUS AND SPECIES	***Leiopelma hochstetteri***

LENGTH
Up to 2 in. (5 cm)

DISTINCTIVE FEATURES
Most primitive of frogs. Very little webbing on feet; large, rounded or pearshaped tongue; degenerate ears, lacking both eardrum and eustachian tube.

DIET
Insects and their larvae; spiders; wood lice (sow bugs)

BREEDING
Breeding season: winter months; number of eggs: 2 to 11 per clutch; hatching period: not known

LIFE SPAN
Not known

HABITAT
Near mountaintops, on ground close to small streams and water seepages; water usually less than 40° F (4° C)

DISTRIBUTION
New Zealand: throughout North Island and on northernmost tip of South Island

STATUS
Thought to be very rare

Hochstetter's frog

in the northwest and in the far southwest of Canada. The male's apparent tail is an organ used for internal fertilization. External fertilization, which is more usual in frogs, would be impossible in the fast-flowing mountain streams where these frogs live, as the sperm would be washed away.

Breeding time for tailed frogs occurs from May to September. The females lay their eggs in strings of 30 to 50, which are fastened to the undersides of rocks. The eggs hatch a month later and within an hour the blackish tadpoles, up to 2 inches (5 cm) long, grow a triangular, adhesive disc on their heads. Using this they cling to rocks, so avoiding being swept away by the current. The tadpole is peculiar in having funnel-like nostrils that it can close when the rush of water is too great. The tadpoles scrape small algae from the rocks for food while holding on with the sucker and can also extract suspended food particles. They turn into froglets between July and September, up to 3 years after hatching.

No voice, no ears

The males of typical frogs croak and all frogs have ears, except for Hochstetter's frog and its relatives. The latter species have what appear to be degenerate ears. They lack both eardrum and eustachian tube (the tube that connects the inner ear with the throat). It has been suggested that American tailed frogs are voiceless and at least semideaf because in the turbulent streams in which they live the females would not hear the males calling above the noise of the gushing water. As a result, the males have lost their voices. Since the frogs are voiceless, there is no use for ears, and these have degenerated.

Hamilton's frog is one of only four species in the Leiopelmatidae family. Like its close relative, Hochstetter's frog, it is found in the mountains of New Zealand, near small streams and water seepages.

HOGNOSE SNAKE

Although they are found in a variety of habitats, hognose snakes of all three species are especially common in areas with sandy or dry soils. They may also be found in open woodlands and areas of scrub, dry riverbeds and orchards. Where the ranges of the species overlap, they may be distinguished by the underside of the tail. This is black in the western hognose and mottled gray or green-gray on a yellow or pink background (the same color and pattern as the belly) in the southern hognose. In the eastern species, the underside of the tail is lighter than the belly.

Hunt frogs and toads

The snout of these snakes is sometimes used to help them to push their way through dense vegetation and loose, sandy soil. Toads, which are often very common in places with sandy soils, form a major part of their diet. Adult hognose snakes have also been recorded as feeding on lizards, fish, birds and small rodents. It is probable that the juvenile snakes also eat any earthworms that they may come across as they push their way through the soil.

As with most snakes, prey is swallowed headfirst. In the case of hognose snakes this process is assisted by long, fanglike teeth at the back of the jaws. It has been suggested that the snakes use these teeth to puncture the skins of toads that have inflated themselves in self-defense, but there are few reliable records of this being observed. Eastern hognose snakes more than the other species rely on toads as food, and in places where toads are becoming uncommon because of land drainage or pollution, the snakes are also decreasing in numbers.

Eggs swell before hatching

The female hognose snake lays her eggs in a damp place, such as under a rotting log or under a stone, in June or July. Each female lays 12 to 30 white, leathery eggs, occasionally more. As they develop, the eggs swell until, just before hatching, they are nearly spherical and have increased in volume by one-third. Newly hatched eastern hognose snakes measure 6–8 inches (15–20 cm) in length. The young are gray, rather than the brownish color of their parents, but they do have the same rows of dark markings.

Hognose snakes are easy to identify due to their unusual upturned snouts, but separating the three species is much more difficult. The western hognose snake (above) is the only one with a black underside to its tail.

THERE ARE THREE SPECIES OF hognose snakes in North America. These nonpoisonous snakes receive their name because all of them have a sharply upturned tip to the snout, rather like farmyard hogs. They all have short, broad heads, a thick body, scales with raised, longitudinal keels and short tails. The background color of the body is variable. It is most often olive green, brown, gray or slate colored. There are usually rows of large dark blotches on the back and flanks. Occasionally uniform black or dark brown specimens are seen.

The eastern hognose snake grows to the largest size, up to 48 inches (1.2 m) in total length, although most adults are within the range 18–30 inches (45–75 cm). These snakes are found from Ontario to the south of Florida, and from the eastern seaboard of the United States to the western borders of Oklahoma and Kansas. Western hognose snakes are rarely longer than 25 inches (63 cm). Their distribution extends in a band from Alberta and Saskatchewan to New Mexico, Texas and northern Mexico. The smallest species is the southern hognose snake. The largest specimen ever recorded was 25 inches (64 cm) in length, but most adults of this species are less than 20 inches (51 cm).

HOGNOSE SNAKES

CLASS **Reptilia**

ORDER **Squamata**

SUBORDER **Serpentes**

FAMILY **Colubridae**

GENUS AND SPECIES **Eastern hognose snake,** *Heterodon platyrhinos;* **western hognose snake,** *H. nasicus;* **southern hognose snake,** *H. simus*

ALTERNATIVE NAMES
Hissing sandsnake; blow viper; puff adder; spread adder

LENGTH
Eastern hognose snake: 18–30 in. (45–75 cm). Western hognose snake: up to 25 in. (63 cm). Southern hognose snake: 20 in. (51 cm).

DISTINCTIVE FEATURES
Fairly thick body with short tail; blunt, broad head; upturned tip to snout; often olive green, brown, gray or slate colored, with dark markings

DIET
Mainly toads with some other amphibians; occasionally lizards, small rodents and birds

BREEDING
Breeding season: eggs laid June–July; number of eggs: 12 to 30

LIFE SPAN
Not known

HABITAT
Mainly sunny places with dry, sandy soil, such as open woodlands, areas of scrub, dry riverbeds and orchards

DISTRIBUTION
South-central Canada; eastern and central U.S.; northern Mexico

STATUS
Generally common but numbers decreasing

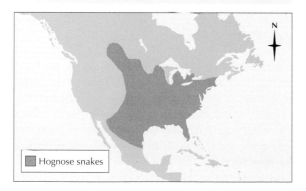

Hognose snakes

Many lines of defense

Nonvenomous snakes like the hognose have many lines of defense. Some, such as the European grass snake, *Natrix natrix,* may feign death in a convincing way. Others, such as the African egg-eating snake, *Dasypeltis scabra,* may inflate the body and hiss, and may even rear up into the kind of strike posture that a venomous snake might adopt. In addition, many kinds of harmless snakes have a superficial resemblance to venomous species, for example the North American coral snakes.

Hognose snakes do all of these things. Many of them look rather like massauga rattlesnakes, *Sistrurus catenatus,* and if approached, will often inflate the front part of their body by drawing air into the lungs, rear up in a most threatening way and hiss loudly. They may also strike, but the mouth is always closed. Hognose snakes can also spread their ribs so that the coloration of the front part of their body is prominently displayed. If all of this fails, they will change tactics and feign death, rolling over onto their backs, opening their mouths and letting their tongues hang limp. They may even give a realistic-looking death twitch. Once the animal has entered this state, it will continue to turn over onto its back however many times it is returned to its normal position.

Killed by mistake

Because they are such apparently aggressive animals, hognose snakes have many local names, including blow viper, puff adder, spread adder and hissing sandsnake. They are frequently killed on sight by people who do not know their true identity. Although still common in most of their range, hognose snakes are decreasing in numbers as a result of this persecution.

Hognose snakes rely on frogs and toads for food. In the above photograph an eastern hognose is swallowing a pickerel frog in the customary manner: whole and headfirst.

HONEY ANT

Honey ants have long been prized as tasty delicacies by people in some parts of the world. Aboriginal Australians, for example, hold the ants' thorax between finger and thumb before nipping off the honey-filled abdomen with their teeth.

ALSO CALLED HONEY-POT ants, these are species that live in dry or desert regions, in which some of the workers remain in the nest and act as living storage vessels. They are then known as "repletes." The habit has been developed independently in various groups of ants belonging to two subfamilies, the Camponotinae and Dolichoderinae, living in North America, Australia and Africa. The so-called "honey" is a sugary solution obtained by the ants from aphids and, in America, from the secretion of a gall growing on small oak trees. The use of repletes is important in deserts because normal food sources are not available during long periods of drought. A kind of halfway condition is seen in the common American ant *Prenolepis imparis*. It feeds largely on honeydew from aphids, and the workers have unusually distensible abdomens, which are often seen swollen to a "semireplete" condition. This probably represents a stage in the evolution of the fully developed honey ants.

More take than give

In the nest of the honey ants some of the workers fail to go out foraging with the rest, remaining at home from the time they leave the pupa. They are perfectly normal ants, at first differing in no way from the workers that hunt for food and perform ordinary duties in the nest, but their behavior is peculiar. Ants constantly feed each other mouth-to-mouth, and these individuals accept food from incoming workers far beyond their own needs. They also give food to others when it is solicited, but when plenty of food is available they take far more than they give.

As a consequence of this excessive intake, the abdomen of these ants becomes more and more distended, taking the form of a globe ¼–⅓ inches (6–8 mm) in diameter. It is translucent with narrow black bars, which are the body segments that were in contact with each other when the ant had its normal shape. When fully replete they hang from the roof of the deeper chambers of the nest. If one of them falls it cannot move, as its fantastically swollen stomach is far too heavy, even if it happens to land in a position from which its feet can reach the ground. If its overloaded crop splits and spills its burden, the other workers rush to enjoy the feast, wholly disregarding the fate of their crippled relative.

Dies to feed others

While there is green vegetation around and food is plentiful, the hanging repletes are constantly visited by incoming ants and persuaded to add more and more to their store. In time of drought, when the foragers return empty or cease trying to find food in the parched and sterile desert, visitors to the repletes solicit the mixed food and

HONEY ANTS

PHYLUM	**Arthropoda**
CLASS	**Insecta**
ORDER	**Hymenoptera**
FAMILY	**Formicidae**
SUBFAMILY	**Camponotinae and Dolichoderinae**
GENUS	***Myrmecocystus* (Americas); *Plagiolepis* (Africa); *Melophorus*, *Leptomyrmex*, *Camponotus* (Australia)**
SPECIES	***Myrmecocystus mexicanus*; *Plagiolepis trimeni*; *Melophorus bagoti*; *Camponotus inflatus*; others**

ALTERNATIVE NAME
Honey-pot ant

LENGTH
½–1 in. (1–2 cm)

DISTINCTIVE FEATURES
Stomach of some workers greatly distended

DIET
Mainly other arthropods such as termites; when these scarce, feed on sugary solution ("honey") stored in some workers' stomachs

BREEDING
Generally as other ants; type of worker determined by age or diet, depending on species

LIFE SPAN
Queen: several years. Workers: much less.

HABITAT
Dry regions, including deserts

DISTRIBTION
Arid regions of western U.S., Mexico, Australia, New Guinea, New Caledonia and southern Africa

STATUS
Abundant in parts of range

A honey ant of the species Melophorus bagoti, *its abdomen hugely distended by stored honeydew.*

she begins to swell and climbs for comfort to the roof of the chamber. Her fate is then sealed; instead of taking her place as an active, busy member of the community, this honey ant must spend the rest of her life as an inert, swollen barrel of syrup.

The honey ants found in the deserts of Arizona have quite sophisticated foraging behavior. The workers prey on insects, particularly termites. When a scout honey ant meets a party of termites it returns to its nest, laying down a trail as it goes. Other workers follow this trail to the termites and carry them back to the honey ant nest.

Honey ants were first discovered in 1881 by an American cleric named Henry C. McCook. The scene of their discovery was the Garden of the Gods in Colorado, and the classically minded McCook gave the ant the specific name *hortideorum*, Latin for the romantically sounding name of its home territory. It is now regarded as a subspecies of *Myrmecocystus mexicanus.*

Ants on the menu

The country people of Mexico search eagerly for the nests of these ants and regard the swollen repletes as a gastronomic delicacy. It is not easy work digging them out as the ants nest on dry ridges where the soil is very hard, but the reward is worth the effort since a well-stocked nest may contain 50 repletes and sometimes as many as 300 may be found. The American entomologist Dr. Alexander Klots writes of the repletes, massed along the ceiling of a horizontal gallery, as gleaming like amber beads in the rays of a flashlight. He describes their contents as "extremely sweet and delicately flavored, far surpassing, in our opinion, honeybee honey."

water, and the swollen bellies of the repletes gradually diminish. They can never return to a normal existence, however, as the stretched skin of their abdomens cannot contract. They probably die as soon as their store is exhausted.

When supplies are coming in and all the established repletes are distended to capacity, any young worker ant may accept a proffered drop from an incoming worker and then find herself besieged by more and more of them until

HONEY BADGER

THE HONEY BADGER resembles the Eurasian and American badgers in build, but is not particularly closely related to them. Like those badgers it is short-legged with a heavy body and a short tail. The honey badger's neck and shoulders are extremely powerful and the forefeet bear claws longer and stouter than those of a Eurasian badger. The head and body of a honey badger total 2–2⅔ feet (60–80 cm) in length with a tail that is 8 inches to 1 foot (20–30 cm) long. The honey badger stands about 1 foot (30 cm) high at the shoulder. The underparts, sides of its body and face are usually dark brown or black in color, while the top of its head, neck and back are light gray or white. This coloration makes the honey badger particularly conspicuous in daylight. Some honey badgers, especially in the Ituri Forest of the Democratic Republic of Congo (formerly Zaire), are wholly black.

Several of the family Mustelidae have striking black-and-white patterns, including polecats, skunks, African weasels and badgers. In some cases at least it seems that this is warning coloration, telling predators that such animals are best left alone. The skunks, with their habit of squirting a powerful and irritating liquid at predators, are a good example. The colors of the honey badger appear to serve a similar purpose, given that it is an aggressive animal and can secrete a foul-smelling fluid from its tail glands.

The honey badger exhibits an unusual mammalian coloration, being lighter on the upperparts of its body and head and dark on its belly, legs and face.

Honey badgers are found in most of Africa from Senegal and the Sudan in the north down to Cape Province in South Africa. They also range across Asia from Arabia and Turkestan to India. They are common across most of their African range, but are becoming rarer elsewhere in the face of advancing human settlement.

Nocturnal prowlers

Honey badgers live in many kinds of country, including rocky hills, forests, savanna, waterless plains and farmland. Even where populations are widespread, they are not often seen because of their nocturnal habits. During the day honey badgers lie up under rocks or in abandoned aardvark holes. If these are not available, they will dig their own burrows. They roam about singly or in pairs, except when the cubs are taken out by their parents.

Eat almost anything

Honey badgers are generalist predators and scavengers, and almost any kind of food is acceptable to them. They eat fruits and berries, honey, insects and all sorts of small and medium-sized vertebrate prey and carrion. Tortoises are eaten after their shells have been smashed by the honey badger's strong teeth and large snakes, even pythons, may be tackled. Their thick skin makes honey badgers almost immune to venomous snakes such as cobras, which they also catch. Honey badgers specialize in the excavation of prey and they often dig up the burrows of ground squirrels, meerkats (suricates), rats and mice, and eat the occupants. They will sometimes attack porcupines, being protected from the porcupines' quills by their tough skin.

Led to bees' nests

Honey badgers have developed an association with the greater honeyguide, *Indicator indicator*. In tropical Africa this bird attracts the attention of honey badgers, humans and some other animals with a special call. It then leads them to bees' nests and the animal breaks open the nest to get to the honey inside. While it is eating, the honeyguide feeds on the wax of the combs that are scattered about. Once again, the honey badger's tough skin protects it, this time from the stings of the bees.

Increasingly, honey badgers themselves are hunted by humans because they occasionally kill livestock such as sheep and chickens, but also because they destroy beehives. Trapping and poisoning has more or less eradicated the honey badger from some areas.

HONEY BADGER

CLASS	**Mammalia**
ORDER	**Carnivora**
FAMILY	**Mustelidae**
GENUS AND SPECIES	***Mellivora capensis***

ALTERNATIVE NAME
Ratel

WEIGHT
15½–28⅔ lb. (7–13 kg)

LENGTH
**Head and body: 2–2⅔ ft. (60–80 cm);
shoulder height: about 1 ft. (30 cm);
tail: 8–12 in. (20–30 cm)**

DISTINCTIVE FEATURES
**Badgerlike with powerful build, particularly
shoulders and neck; flat, blunt head; small
ears; light gray or almost white back, top of
head and neck; black face and underparts;
long, powerful claws**

DIET
**All small and medium-sized vertebrate prey
and carrion; also invertebrates, fruits,
berries and honey**

BREEDING
**Age at first breeding: not known; breeding
season: varies according to region; gestation
period: 150–180 days; number of young:
usually 2; breeding interval: not known**

LIFE SPAN
Up to 26 years in captivity

HABITAT
**Mainly savanna, scrub and desert; also
forests and farmland**

DISTRIBUTION
Africa, east through Middle East to India

STATUS
Common in most of Africa; rarer elsewhere

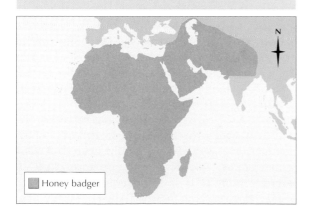

Honey badger

Breeding little known

Very little has been recorded about the breeding habits of honey badgers. Most activities take place at night, and their ferocious disposition has in the past discouraged close investigation. They have large, overlapping ranges and are normally seen foraging singly or in pairs. It has been suggested that such pairs might represent long-term mates, although short-term liaisons are more likely. The breeding season depends on the latitude at which the honey badgers live. The young are born after a gestation period of around 5–6 months and there usually two per litter, although the range is from one to four cubs. The young are reared in lined burrows or in deep crevices in rocks. Once weaned, the cubs will stay with their mother for an unknown period before becoming independent.

Formidable foe

The honey badger is one of the most ferocious animals for its size and will sometimes attack without provocation, especially during the breeding season. Its success as a fighter is not just because of its long claws and strong teeth. Any predator that can get past these weapons finds a honey badger difficult to kill. Its skin, impervious to bee sting and snakebite, is exceptionally tough and has been described as being "like a loose coating of rubber." If an opponent grabs it by the scruff of its neck, the honey badger can turn around inside its skin and deliver a severe bite. There are records of honey badgers engaging a pack of dogs and, after a long and ferocious fight, trotting away leaving the dogs exhausted and wounded.

*Although still common
across much of their
range, honey badgers
are often trapped or
poisoned because they
destroy beehives and
sometimes take
livestock.*

HONEYBEE

In the colonies of social bees there are two kinds of females. The fertile females are called queens and the sterile females are the workers. The workers do all the work of maintaining the economy of the colony. The male honeybees are called drones.

Giant relatives

Only four species of the genus *Apis* are known, and one of them, the eastern honeybee, *A. indica*, is so similar to the European honeybee that it is sometimes regarded as a subspecies. The eastern honeybee is domesticated in tropical Asia. Both the other species inhabit the eastern Tropics. The giant honeybee, *A. dorsata*, is a large bee that makes enormous hanging combs in the open. Large branches overhanging cliffs and buildings, especially water towers, are favorite sites for colonies. These bees may be dangerous if disturbed. Despite this, a tribe called the Dyaks of Borneo climb by night, throw down the combs and gather the honey. The small honeybee, *A. florea*, is by contrast a fairly docile insect. A colony consists of a single comb that contains only 1 or 2 ounces (28–56 g) of honey.

Household chores

The great majority of European honeybees are now living in hives, although wild colonies may be found, almost always in hollow trees. In midsummer a strong colony normally contains one queen, 50,000 to 60,000 workers and a few hundred males or drones.

Worker bees are short lived, surviving perhaps 4 to 6 weeks. For the first 3 weeks after emerging from the pupa the worker's duties lie within the hive. Her first spell of work is as nursemaid to the developing larvae, and she passes on a great deal of the food she eats to these, partly by direct regurgitation and partly by giving them a jellylike secretion from certain salivary glands on her head. By the time she is around 12 days old her wax glands have developed and she turns to building and repairing the comb of geometrically arranged cells in which the larvae are reared and food is stored. At this time she also starts to take the nectar and pollen brought in by returning foragers, converting the nectar to honey and storing it away. At the same

Honeybee workers on a honey and pollen storage comb. The female workers maintain the colony: they feed the larvae, build and repair the comb and forage for nectar, pollen, water and resin.

ANY OF THE FOUR SPECIES of social bees belonging to the genus *Apis* can be called honeybees. However, the name is most usually associated with the European domestic bee, *A. mellifera*, sometimes called the western honeybee. This species differs from all other social bees and social wasps of temperate climates in that it forms colonies that survive the winter by living on reserve stores of food. As a result, European honeybees may occupy a particular dwelling site or nest for an indefinite length of time. In social wasps and bumblebees, on the other hand, all the members of the colony die at the end of the summer with the exception of the fertilized females or queens. These hibernate and found new colonies the following spring.

time she helps to keep the hive tidy, carrying dead bees and other debris outside. At 3 weeks old the worker bee is ready to go out foraging for nectar, pollen, water and resin, which are the four substances needed for the hive's economy.

Resin is used to make a sort of varnishlike cement called "propolis" with which any small openings or crevices in the hive are sealed.

Searching for nectar

In searching for nectar-yielding flowers, the worker bee is guided by her senses of smell and sight. Honeybees cannot see red at all but can see ultraviolet light, which is invisible to humans. Bees guide themselves to and from the hive by reference to the angle of the sun, or to the angle of polarized light from the sky. They have a sense of time that enables them to compensate for the continuous change in the sun's position.

After 2 or 3 weeks of foraging the worker is worn out and dies. Workers hatched in the autumn have a longer life before them, as they build up food reserves in their bodies and survive through the winter. During this time their activity is greatly reduced. The bees keep warm by huddling together in a mass and feeding on the honey that they have stored.

The queen rules the great horde of female workers by secreting a substance from her body, the presence or absence of which controls their behavior. Her chief role, however, is egg-laying, and at midsummer she may be laying 1,500 eggs a day. This enormous fecundity is needed to compensate for the shortness of the workers' lives.

Mating with and fertilizing the queens is the only useful part played in honeybee economy by the male drones. Males that do mate with the queen die immediately after. During summer

HONEYBEES

PHYLUM	**Arthropoda**
CLASS	**Insecta**
ORDER	**Hymenoptera**
SUBORDER	**Apocrita**
FAMILY	**Apidae**

GENUS AND SPECIES **European honeybee,
Apis mellifera; eastern honeybee, *A. indica*;
giant honeybee, *A. dorsata*; small honeybee,
*A. florea***

ALTERNATIVE NAMES
***A. mellifera*: hive bee; western honeybee**

LENGTH
**Queen (fertile female): ⅗ in. (1.5 cm).
Worker (sterile female): ⅖ in. (1 cm).
Drone (male): slightly larger.**

DISTINCTIVE FEATURES
Social bees that build complex nests

DIET
**Adult bees: flower pollen and nectar.
Worker and drone larvae: nectar, honey
and pollen; some royal jelly (secretion
produced by workers' special glands).
Queen larva: royal jelly alone.**

BREEDING
**Breeding season: all summer; number of
eggs: up to 1,500 each day; hatching period:
3 days; larval period: 6 days (worker),
8 days (drone); pupal period: 12 days**

LIFE SPAN
**Queen: several years. Worker: usually 4–6
weeks; ones that develop later in year will
survive winter. Drone: 4–5 weeks.**

HABITAT
**Domesticated bees (majority): nest in hives.
Wild bees: nest in hollow trees.**

DISTRIBUTION
**Probably originated in warmer regions,
such as Southeast Asia. *A. mellifera*: reared
domestically in most western countries.
Other species: eastern Tropics.**

STATUS
Common

Foraging for nectar and pollen is hard work. After just 2 or 3 weeks of foraging, worker bees become worn out and die.

Fertilized and unfertilized eggs

Queens may be produced in a hive in response to the urge to swarm or because of the aging of the mother queen. In either case the queens fly out to seek mates immediately on becoming adult. The sperm is stored by the queen in an internal sac called the spermatheca, and sperm is released to fertilize the eggs as she lays them. All eggs that are fertilized produce females, either workers or queens. Drones are only produced from eggs that develop without being fertilized.

The larval and pupal stages of honeybees (collectively known as the brood) are passed in the wax cells into which the eggs are laid, one in each cell. The larvae are entirely helpless and are fed by the workers. The development of a worker bee takes around 3 weeks: 3 days as an egg, 6 as a larva and 12 as a pupa. The natural mating behavior of queen and drone bees makes any control of pairing and breeding impossible, but in recent years a technique for artificially inseminating chosen queens with sperm from chosen drones has been developed.

Common and royal food

The natural food of bees consists of nectar and pollen. The bees also make honey from nectar and store it for food. The larvae are fed partly on a mixture of nectar or honey and pollen, and partly on a secretion from various glands of the workers, the substance that is often called royal jelly. When a fertilized egg is laid in a normal-sized cell, the larva is fed at first on jelly and later on pollen and honey, and it develops into a worker. When production of queens is needed, the workers make larger cells into which the queen lays ordinary fertilized eggs. The larvae from these, however, are fed on royal jelly alone and they develop into queens. Drone larvae are fed similarly to those of workers but for a few more days and in slightly larger cells.

Predators and disease

In spite of their stings, bees are preyed on by birds, dragonflies and some kinds of wasps. Wax moths lay their eggs in the hives and the larvae live on wax, pollen and general comb debris, sometimes doing serious damage. The large death's head hawkmoth, *Acherontia atropos*, invades colonies and steals the honey, piercing the wax comb with its short, stiff proboscis. The greatest threat to honeybees, however, is disease and starvation.

Cells cut away to show pupae. Females are produced from fertilized eggs, while the male drones are produced by unfertilized eggs. To produce queens, the female larvae are fed on royal jelly alone.

they usually live for 4 or 5 weeks and are fed by the workers. In autumn any drones remaining in the colony are driven outside to die.

Swarming

New colonies are founded by what is known as swarming. At this time extra queens are produced in the hive and then large numbers of workers, accompanied by some drones and usually one queen, leave the hive and fly together for some distance. They then settle in a large cluster and search for a suitable site for the colony. Once one is found, the new colony is made by some of the workers. At this stage they can easily be persuaded to settle down in artificial quarters of any kind merely by shaking the swarm, with its attendant queen, into a suitable receptacle.

Honeybees have been kept by humans for their honey for many hundreds of years. This has been mainly a matter of inducing the bees to make colonies in hollow receptacles of various kinds, such as earthenware pots, logs and straw baskets, and then taking the honey they produce.

HONEY BUZZARD

HONEY BUZZARDS BELONG TO the same family, Accipitridae, as the hawks and eagles. They are so named because they feed largely on bees and wasps, and in flight they have a silhouette very much like that of a buzzard. However, honey buzzards make up the genus *Pernis*, whereas the true buzzards are members of the genus *Buteo*. In the United States the latter are named hawks, and the name buzzard is popularly used to describe vultures.

The wing feathers of honey buzzards are splayed at the tips. However, compared to true buzzards the honey buzzards' wings are, on the whole, longer and narrower, and their heads and necks are also longer. The upperparts are dark brown, and the underparts range from white with a few dark markings to a dark brown similar to the that of the upperparts.

The Eurasian honey buzzard, *Pernis apivorus*, breeds in most of Europe and in Asia as far east as Lake Balkash in Kazakhstan and the River Ob in Siberia, but is missing from the greater part of Norway and northern Scandinavia. The related crested honey buzzard, *P. ptilorhynchus*, lives in India, China, eastern Siberia, parts of Japan, Southeast Asia and the surrounding islands. In the Celebes, off Sulawesi, and in the Philippines there is the barred honey buzzard, *P. celebensis*. The long-tailed honey buzzard, *Henicopernis longicauda*, lives in New Guinea and on neighboring islands.

Crowded commuters

Honey buzzards live in woodland where there are plenty of large trees and are often found in clearings or along more open ground by the side of roads and streams. At the end of their breeding season the Eurasian honey buzzards that breed in Europe migrate to West, central and southern Africa, while those breeding farther east move south to India and Southeast Asia.

Migrating honey buzzards often travel in large numbers, not so much in flocks as in continuous streams following well-defined routes, along the coasts of lakes, for example. They cross the Mediterranean at Gibraltar or other places where the sea crossing is as short as possible, swarming across when the weather is favorable. Tens of thousands of birds cross the Strait of Gibraltar or the Bosporus on their way to Africa.

Concentrate on wasps

"Wasp buzzard" would be a better name, as a honey buzzard's main food is the grubs of wasps. It does occasionally eat honey and wax, but the name of honey buzzard was probably given because it was often seen digging out the combs from the large nests of wasps and honeybees, as well as the smaller nests of bumblebees. The honey buzzard's main interest in these combs is the larvae and pupae inside, along with the cell walls which provide roughage, presumably absent elsewhere in its high-protein diet.

Rather than soaring high in the air, honey buzzards are more often seen on the ground, where they move nimbly about. Having detected a suitable nest, they dig it out, scraping the earth back with both feet and pulling out pieces of comb. Sometimes quite a deep hole is necessary. Honey buzzards also feed on ants, moths, cockchafers and locusts, and they occasionally take small mammals, slugs, snakes, earthworms, frogs and the eggs and nestlings of other birds. They also eat some fruit.

Decorating the nest

Courtship is like that of true buteo buzzards and involves much swooping and soaring. At the top of the soar the honey buzzard hovers for a few seconds and claps its wings over its back two or

A honey buzzard with its young. Male and female alike decorate the nest with leaves and greenery, both before and after the chicks are hatched.

<table>
<tr><td colspan="2">HONEY BUZZARD</td></tr>
</table>

CLASS	**Aves**
ORDER	**Falconiformes**
FAMILY	**Accipitridae**
GENUS AND SPECIES	***Pernis apivorus***

ALTERNATIVE NAME
Eurasian honey buzzard

LENGTH
Head to tail: 1¾–2 ft. (52–60 cm); wingspan: 4½–5 ft. (1.35–1.5 m)

DISTINCTIVE FEATURES
Longish wings with fingered primary (flight) feathers; long tail; head and neck protrude more than in true buzzards of genus *Buteo*; dark brown upperparts; variable underparts, from buff with dark streaks to entirely dark

DIET
Mainly larvae, pupae and adults of wasps, hornets, honeybees and bumblebees; also other insects, slugs, earthworms, small vertebrates, nestlings, bird eggs and fruits

BREEDING
Age at first breeding: probably 3 years; breeding season: eggs laid May–early July; number of eggs: 2; incubation period: 30–35 days; fledging period: 40–44 days; breeding interval: 1 year

LIFE SPAN
Up to 30 years

HABITAT
Usually woodlands with clearings; also mixed woods, copses and meadows

DISTRIBUTION
Summer: northern Spain east across most of Europe to western Siberia and Iran. Winter: mainly western and central Africa.

STATUS
Locally common

Honey buzzard (summer)

Despite its buzzardlike appearance in flight, with the wing feathers splayed at the tips, the honey buzzard has longer, narrower wings than the true buzzards of the genus Buteo.

three times in quick succession. Both the male and the female also soar high over the nest, the male diving at the female.

The honey buzzard pair usually builds its nest on the abandoned nest of a crow or buzzard in a tall tree, beech trees being preferred. Both sexes decorate the nest, bringing fresh sprigs of greenery each evening to line it. When the chicks have hatched, they squeal when the sprays of leaves are brought in and often play with them. The usual clutch is of two eggs, which are incubated for up to 35 days, mainly by the female. The chicks are fed at first on wasp grubs, which are regurgitated to them, or else the male brings pieces of comb and the female picks out the grubs for the chicks. Later frogs are brought and, until the chicks learn how to eat them, these are skinned by the parents. The chicks fly when 40 to 44 days old. Until they can balance on one leg properly and hold a piece of comb in the foot, the chicks return to the nest to feed. They also return to the nest to roost every night.

Avoiding the sting
Using much the same method as bee-eaters, honey buzzards eat bees and wasps by first carefully removing their stings. They catch a wasp using the bill, seizing it around the middle, and then rip the sting off before the wasp is swallowed. Their main food, however, is not the adult insects but the defenseless grubs, which do not need such treatment.

HONEYCREEPER

THERE ARE TWO GROUPS of birds called honeycreepers, comprising 15 species. One of these, Emberizidae, consists of birds related to tanagers and wood warblers that have developed the habit of nectar eating. The second group is that of the Hawaiian honeycreepers, Drepanididae, birds that invaded the islands of Hawaii and developed a variety of forms. Many members of the first group are known as sugarbirds, which can be confusing because some of the birds known as honeyeaters (discussed elsewhere) also have this name.

Honeycreepers are small with thin bills and the males are usually brilliantly colored, often with bright shades of blue plumage. The females have duller, usually greenish plumage. The green honeycreeper, *Chlorophanes spiza*, which is 5 inches (13 cm) long, is one of the largest species. The male's plumage shines bright green to turquoise in color, depending on subspecies, of which there are seven. It has a yellow bill, black head and red irises to its eyes. The female is green with no black on the head. Several species of dacnis, also called honeycreepers, are closely related. The male blue dacnis, *Dacnis cayana*, for example, is blue with black on much of its head, back, tail and wings. The female is grass green in color.

Honeycreepers live in the American Tropics, from Mexico and Central America through most of South America, and in the Caribbean. The green honeycreeper, blue dacnis and red-legged or blue honeycreeper, *Cyanerpes cyaneus*, are the three most widespread species. Hawaiian honeycreepers, meanwhile, are confined to the Hawaiian archipelago and have a variety of local names, such as *palik* and *akiapolaau*.

Honeycreepers usually live in flocks in forests or plantations, sometimes mixing with tanagers and other birds. Some honeycreepers are quite aggressive. The green honeycreeper, for example, seizes other small birds that are competing for the same food. Male green honeycreepers have also been seen to attack their mates while nest-building. The female scarlet-thighed dacnis, *Dacnis venusta*, is known to attack larger, but harmless species such as thrushes.

Nectar-eaters

The honeycreepers are probably a miscellaneous collection of birds that are united by convergent evolution in feeding on the same food: nectar. Their tongues have brushlike tips similar to those of other nectar-eaters, such as the honeyeaters. This brush helps to sweep up the nectar. The

tongue is also modified into a tubular structure which, together with the thin bill, is adapted for sucking nectar. Most honeycreepers also feed on fruits and take some insects.

Honeycreepers usually live in pairs and build cup-shaped nests in dense foliage in trees and shrubs. The nest of the scarlet-thighed dacnis, for example, is a frail hammock, camouflaged from below by pieces of green fern. In the green honeycreeper, egg laying takes place from April to July. The female usually lays two eggs, which she incubates for some 13 days.

Hawaiian honeycreepers

The Hawaiian honeycreepers appear to be descended from an ancestor, perhaps a nectar-drinking tanager, that arrived from America 2,000 miles (3,200 km) away, presumably blown off course by the wind. From this ancestral type around 23 species evolved. They are between 4 and 8 inches (10–20 cm) long, and their colors range from green, gray and black to bright red and yellow. Their breeding habits are fairly uniform. Simple nests are built in trees, bushes or long grass and two or three eggs are laid in them. Normally the female bird incubates the eggs and is fed away from the nest by the male.

The diversity of the Hawaiian honeycreepers really lies in their bills and feeding habits. The subfamily, Drepanidinae, have curved bills and brush-tipped tongues for nectar-drinking and probing for insects. There are variations on these

Honeycreepers have adapted to eating nectar but, like their close relatives the tanagers, also feed on insects and fruits. The green honeycreeper is pictured.

A male purple honey-creeper, Cyanerpes caeruleus, *at the Asa Wright Nature Reserve, Trinidad.*

bill patterns. Some in the family use their long bills for probing insects out of crevices in bark, while others chisel away bark by hammering.

No competition

The evolution of different forms and feeding habits of the Hawaiian honeycreeper was aided, like the adaptive radiation (evolutionary diversification) of Darwin's finches of the Galapagos Islands, by the absence of competition from other birds. Changes in both these birds are in feeding rather than in breeding habits because the environment for breeding is uniform on both the Galapagos and Hawaii. On the other hand, there is a diversity of available food. The diet of different Hawaiian honeycreepers includes nectar, seeds, fruits, spiders and insects.

The first stage in the evolution of the Hawaiian honeycreepers is a diversification of nectar-drinking types. In the Tropics nectar-drinkers evolve faster than insect-eaters. Insects are everywhere, but nectar is abundant only locally, so the nectar-drinkers have to travel in search of it. In this way they move to new places, where some may become isolated and are able to evolve in a different way from the original stock. They may be isolated on real islands or on environmental "islands" such as mountaintops.

Many endangered or extinct

In the 19th century nearly a half of the various species of Hawaiian honeycreepers became extinct, and many others are now endangered. At one time their feathers were used by the Hawaiians in the manufacture of cloaks and it is often said that this caused the extinction of some species. Nonetheless, the downfall of the honeycreepers has been more the result of American colonization. Destruction of natural forest habitats, introduced predators, such as cats and rats,

GREEN HONEYCREEPER

CLASS	**Aves**
ORDER	**Passeriformes**
FAMILY	**Emberizidae**
GENUS AND SPECIES	***Chlorophanes spiza***

LENGTH
Head to tail: about 5 in. (13 cm)

DISTINCTIVE FEATURES
Fairly long, slightly decurved bill. Male: plumage shines green to turquoise, with color depending on subspecies; black head; yellow bill; red irises; dark legs. Female: duller green plumage; no black area on head.

DIET
Fruit, berries, insects and nectar

BREEDING
Age at first breeding: not known; breeding season: eggs laid April–July; number of eggs: 2; incubation period: 13 days; fledging period: not known; breeding interval: 1 year

LIFE SPAN
Not known

HABITAT
Flowering and fruit-bearing trees in or near forests or forest edges

DISTRIBUTION
From Mexico south through Central America to South America, including Venezuela, Columbia and Brazil

STATUS
Common

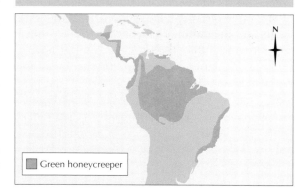

Green honeycreeper

and introduced diseases, such as avian malaria, were problems that the specialized honeycreepers could not compete against. Many of these factors remain a problem today, and it is thought that more species may be lost unless corrective action is taken.

HONEYEATER

HONEYEATERS ARE A diverse group of birds that eat mainly nectar and fruit, but also take insects. They are varied in form and habits. The main family characteristic is the "paintbrush" tongue, adapted to nectar feeding. The plumage in most species is fairly dull, gray brown, greenish or streaked.

The curiously named ooaa or Kauai oo, *Moho braccatus*, of the island of Kauai, Hawaii, searches for insects in tree trunks, propping itself on its stiff tail rather like a woodpecker. The friarbirds, genus *Philemon*, have dull plumage and several, such as the noisy friarbird, *P. corniculatus*, and helmeted friarbird, *P. buceroides*, have horny growths on their bills and vulturelike, bald heads.

The spinebills, genus *Acanthorhynchus*, have long curved bills and hover in front of flowers like hummingbirds. Other honeyeaters resemble tits, flycatchers or warblers. They are generally dull looking and some have wattles and lobes or bare patches of skin on the face.

Confined to Pacific region

More than 60 of the 174 honeyeater species are found in Australia, with a similar number on the island of New Guinea. The remainder are in New Zealand and the islands of the southwestern Pacific. Some range as far east as Hawaii, where they are now less common because of the felling of natural forests. The oo or great oo, *Moho nobilis*, with the shortest name of any bird, is now extinct. Its yellow feathers, sprouting in tufts like epaulettes from under its wings, were used on royal capes. Another Hawaiian species of oo, Bishop's oo, *M. bishopi*, was last seen in 1981. The Maoris used the yellow feathers of the stitchbird, *Notiomystis cincta*, for ornamental capes in the same way as the Hawaiians used the oo. Because of this and the cutting down of forests, it is now very rare and found only on Little Barrier Island off New Zealand. Very few honeyeaters live outside the Australasian region. The brown honeyeater, *Lichmera indistincta*, reaches Bali and another species, *Apalopteron familiaris*, lives on the Bonin Islands off Japan.

Good singers

Many of the honeyeaters are good singers, although not the species that live in dense forests. For example, the yellow-tufted honeyeater, *Lichenostomus melanops*, sings a melodious trill, while the white-gaped honeyeater, *Meliphaga unicolor*, whistles a loud, flutelike song. The tui, *Prosthemadera novaeseelandiae*, is one of the best songsters in New Zealand and is also a good

mimic. It has the alternative name of parson bird because of the two patches of white feathers at the neck, rather like the collar of a clergyman.

Honeyeaters live in a variety of terrain from almost barren country to dense rain forests. The singing honeyeater, *Lichenostomus virescens*, is found in the coastal sand dunes of Australia, but honeyeaters are usually restricted to flowering trees and shrubs and are rarely seen on the ground. Many of them are gregarious, moving around the countryside in groups, sometimes in huge flocks, in search of flowers, berries and fruits. A few species are orchard pests.

It is not just in their plumage that honeyeaters are diverse. On New Guinea their extremes of size are shown by the large helmeted friarbird, which is 13 inches (33 cm) in length, and the pygmy honeyeater, *Oedistoma pygmaeum*, which is only 2⅘ inches (7 cm) long.

Important pollinators

Some honeyeaters live exclusively on nectar. One such bird is the stitchbird, which even feeds its chicks on nothing but nectar. This is unusual because birds normally bring insects for their chicks to provide them with the protein needed for growth. Other honeyeaters have largely lost the nectar-eating habit and do not have the paintbrush tongue. For example, the strong-billed honeyeater, *Melithreptus validirostris*, of Tasmania, behaves like a nuthatch, running up tree trunks and picking up insects from under the bark.

A white-cheeked honeyeater, Phylidonris nigra, *feeding on the plant* Banksia sceptrum. *Most honeyeaters drink flower nectar and play an important role as pollinators.*

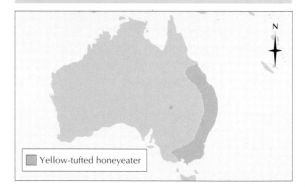

YELLOW-TUFTED HONEYEATER

CLASS	**Aves**
ORDER	**Passeriformes**
FAMILY	**Meliphagidae**
GENUS AND SPECIES	***Lichenostomus melanops***

LENGTH
Head to tail: 6¾–8⅔ in. (17–22 cm)

DISTINCTIVE FEATURES
Longish, slightly decurved black bill; broad, glossy, black mask ending in bright golden ear tufts; yellow crown; dark olive-brown back; clear, golden yellow throat with gray central stripe

DIET
Flower nectar, sap from bark, insects and berries

BREEDING
Age at first breeding: 1 year; breeding season: eggs laid June–December; number of eggs: 2 or 3; incubation period: probably 12–16 days; fledging period: not known

LIFE SPAN
Not known

HABITAT
Forest and woodland with dense undergrowth; also drier scrub

DISTRIBUTION
Australia: Victoria, eastern New South Wales and southeastern Queensland

STATUS
Uncommon

Cardinal honeyeater, **Myzomela cardinalis,** *in Vanuatu, Samoa, South Pacific. Very few honeyeaters are found outside the Australasian region.*

The rufous-throated and rufous-banded honeyeaters (*Conophilia rufogularis* and *C. alisogularis*) also hunt for insects in the same manner as flycatchers, rather than eating nectar.

Honeyeaters have the same specialized digestive system as the nectar-eating flowerpeckers. In this it is possible for food to bypass the gizzard (a muscular enlargement of the alimentary canal), going instead straight into their intestine. In most cases the honeyeater's tongue is also modified for nectar-eating. The tip is divided into four and each part is frayed to form a delicate brush with which the nectar is swept up. Like the flowerpeckers, honeyeaters also pollinate the flowers on which they feed. Together with parrots they are responsible for the pollination of most of the flowering trees and shrubs in Australia.

Nesting high in trees

The honeyeater's nest is cup-shaped, sometimes so loosely woven that the eggs can be seen from underneath. It is usually placed high in a tree. Some of the gregarious species nest in colonies of 20 or so pairs. An unusual habit has been developed by the blue-faced honeyeater, *Entomyzon cyanotis*. It sometimes builds its own nest, but more often it uses the nest of a babbler instead. Sometimes it does not even wait for the nest to be abandoned, but drives the other bird away.

In most honeyeaters, the two or three eggs are incubated for around 12 to 16 days and the chicks are fed by both parents for a similar period. Many species are cooperative breeders, with several adults feeding the young from each nest. Apart from the usual nesting materials such as twigs, grasses and flower stems, some species of honeyeaters incorporate animal hair in their nests, usually as a lining. They do not always wait for the owner to relinquish the hair. The honeyeaters will boldly settle on the animal of their choice and tweak out hair. The natural sources of this hair were native marsupial mammals such as koalas and opossums. Nowadays domestic stock and introduced deer are also raided.

Yellow-tufted honeyeater

HONEYGUIDE

THE HONEYGUIDES ARE a family of small birds that are unusual in both their feeding and breeding habits. They range from sparrow-sized to thrush-sized. The plumage is dull, generally brown or gray above and lighter underneath, but some species have patches of color on the tail, wings or head, and the greater honeyguide, *Indicator indicator*, has yellow wing patches. The most colorful species is the orange-rumped honeyguide, *I. xanthonotus*, with a golden yellow crown and chin and a bright orange rump. The lyre-tailed honeyguide, *Melichneutes robustus*, has black-and-white tail feathers that fan out along the length of the tail, and in the other species the outer two pairs of tail feathers are somewhat shorter than the rest.

There are 17 species of honeyguides, most of them living in Africa south of the Sahara and the others in southern Asia. One relatively uncommon species, the Malaysian honeyguide, *I. archipelagicus*, is found in the rain forests of Thailand, peninsular Malaysia and Borneo.

Beeswax eaters

Honeyguides live singly or in pairs in forest and brush country. They are strong fliers, with a rapid undulating flight. Honeyguides do not migrate, but it seems that they make seasonal movements in search of food.

Honeyguides eat mainly insects, which they may catch on the wing, rather like flycatchers. They will fly out from a perch, catch an insect such as a termite, flying ant or fly, and carry it back to the perch. Locusts are also caught, and honeyguides have been found feeding on fruits and berries, but their main food is bees and wasps, especially wild honeybees; honeyguides are rarely found in places where there are no bees or wasps.

The adult bees are caught on the wing and nests are attacked. Honeyguides have a tough skin, which may be a protection against stings, but there are no records of honeyguides removing the wasp or bee stings before eating them, as do bee-eaters and honey buzzards. It is quite possible that honeyguides are partially immune to stings, because they have been seen disappearing into bees' nests and reappearing apparently unscathed, although the bees within are buzzing furiously.

Honeyguides penetrate bees' nests to eat the larvae and, more particularly, the beeswax from the combs. It was some time before naturalists realized that it was wax rather than honey that attracted honeyguides to bees. Wax-eating is a rare habit in birds, and honeyguides are unique in the amount of wax they eat. It is a very indigestible substance, but honeyguides probably have special enzymes or bacteria in their intestines that break it down.

Honeyguides enlist help

The greater honeyguide is known to guide humans, honey badgers or ratels, *Mellivora capensis*, and other mammals to bees' nests in order to feed on the leftovers after the mammals have torn the nests open. Several mongooses and baboons may also be used in this way. However, naturalists have made relatively few eyewitness reports of this behavior.

Only the two largest species of honeyguides, *I. indicator* and *I. variegatus*, are known to carry out this remarkable behavior, which seems to be started by the birds finding a nest with bees flying in and out. Abandoned nests with plenty of wax are ignored while new nests, occupied but devoid of wax, are attacked. The honeyguide first attracts the attention of its helper by calling and fanning its tail. The call is a rapid churring, like a box of matches being rattled. This call is kept up until the follower moves toward it. Then the honeyguide flies toward the bees' nest and

Honeybees, including the adults, larvae and beeswax, are the preferred food of the greater honeyguide. Unable to gain access to most bees' nests on its own, the bird enlists the help of honey badgers and humans.

The lesser honeyguide, Indicator minor, *has a strong association with honeybees. However, it does not have the greater honeyguide's ability to lure mammals to bees' nests, and can feed only from nests that it can open itself.*

GREATER HONEYGUIDE

CLASS	**Aves**
ORDER	**Piciformes**
FAMILY	**Indicatoridae**
GENUS AND SPECIES	***Indicator indicator***

LENGTH
Head to tail: 7½–8 in. (19–20 cm)

DISTINCTIVE FEATURES
Stout, blunt-tipped bill; feet like those of woodpeckers: 2 front toes and 2 back toes on each foot. Male: dark gray above, pale gray below; large white ear patch; black throat; yellow shoulder patches; pink bill. Female: no white ear patch and black throat.

DIET
Insects, especially honeybees, and beeswax; occasionally fruits; rarely bird eggs

BREEDING
Age at first breeding: 1 year; breeding season: eggs laid September–October; number of eggs: 4 to 8; incubation period: not known; fledging period: 30–40 days; breeding interval: 1 year

LIFE SPAN
Not known

HABITAT
Woodland, forest edge, savanna, orchards

DISTRIBUTION
Much of sub-Saharan Africa, except Horn of Africa and southwest

STATUS
Locally common or uncommon

Greater honeyguide

perches again until its follower catches up. The process is repeated until the nest is reached, and the honeyguide falls silent while the nest is being torn open by the mammal.

The Boran people of Kenya find honey by following greater honeyguides. The average time for a Boran to find a new bees' nest is 9 hours if he or she searches alone, but only 3 hours with the birds' help. About 90 percent of bees' nests are accessible to the birds only after humans have opened them with tools, so both parties gain from the arrangement.

To find out if local honeyguides know of the whereabouts of a bees' nest, the Boran whistle to attract one of the birds. It will then disappear in the direction of the nest if it knows where one is to be found.

Laying in others' nests

The breeding habits of six honeyguides are completely unknown, but the 11 species that have been studied are all parasitic, like certain species of cuckoo. The males sing from special perches within their territories. The females are attracted, mating takes place and the pair splits up, the female departing to find the nest of a suitable host. As well as singing, male lyre-tailed honeyguides perform complex aerial displays and produce beating sounds, caused by the rush of wind vibrating their outer tail feathers.

The female honeyguide lays her eggs in the nests of hole-nesting birds such as barbets, woodpeckers, bee-eaters, hoopoes, woodhoopoes and kingfishers. She waits until the owners are away and then slips into the nest and lays her eggs. At the same time she destroys the hosts' eggs by puncturing them. Any host eggs that are laid after the female honeyguide's visit are able to hatch out, but if this happens the honeyguide chicks kill their nestmates. The upper and lower halves of their bills are armed with sharp hooks with which they repeatedly peck the other chicks until they die.

HOODED SEAL

THE HOODED SEAL GETS its name from the inflatable proboscis, or hood, of the male, which extends from the nostrils to a point just behind the eyes. The function of this hood is not fully understood.

Hooded seals are large animals. The adult males grow to a maximum length of about 9 feet (2.7 m) and can weigh up to 880 pounds (400 kg). Females are smaller, about 7 feet (2.1 m) long, and weigh proportionally less. Adult seals are dark gray in color with a number of irregular, dark markings on the back. These are usually 2–3 inches (5–7.6 cm) across, becoming smaller toward the neck. The patches are often surrounded by a circle of small, whitish specks. The seal is paler on the underside, the female being generally paler all over than the male, with less distinct markings.

Solitary seals

Hooded seals are largely solitary animals, but they congregate in herds at breeding time. There are three main breeding concentrations of hooded seals. These are around the island of Jan Mayen in the Greenland Sea, on the coasts of Newfoundland and in the Davis Strait, between Canada and Greenland. In spite of this comparatively restricted northern range, hooded seals occasionally wander far from their normal home. Several have been seen around the British Isles, one as far south as the Orwell River in Suffolk. Vagrant hooded seals have also turned up in the Bay of Biscay, while on the other side of the Atlantic they have appeared sporadically as far south as Cape Canaveral in Florida.

The world population of hooded seals has been calculated at somewhere between 500,000 and 600,000 individuals, but an accurate estimate is difficult because of their solitary habits. The seals all but disappeared in the 1930s, during a warm climatic period, but they became numerous again during the 1960s, when ice conditions were severe. In more recent years there has been some change in the population, with numbers decreasing around Newfoundland.

Hooded seals are solitary animals except at the breeding time. Mainly found in the Arctic Ocean, individuals do occasionally stray as far south as the Bay of Biscay and Cape Canaveral in Florida.

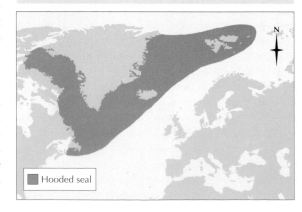

HOODED SEAL

CLASS	**Mammalia**
ORDER	**Pinnipedia**
FAMILY	**Phocidae**
GENUS AND SPECIES	*Cystophora cristata*

ALTERNATIVE NAME
Bladdernose

WEIGHT
**Male: 440–880 lb. (200–400 kg);
female: 320–660 lb. (145–300 kg)**

LENGTH
**Male: 8¼–9 ft. (2.5–2.7 m);
female: 6½–7¼ ft. (2–2.2 m)**

DISTINCTIVE FEATURES
**Silvery gray overall, with black or brown
patches of variable size and shape;
completely black face; prominent, fleshy
proboscis; enlarged nasal cavity creating
an inflatable red sac (male only)**

DIET
Cephalopods, shrimps, mussels and fish

BREEDING
**Age at first breeding: 3 years (female),
13 years or more (male); breeding season:
March–April; number of young: 1; gestation
period: almost 365 days, including delayed
egg implantation; breeding interval: 1 year**

LIFE SPAN
Up to 35 years

HABITAT
Deep waters on and around thick, drifting ice

DISTRIBUTION
**Breeding populations: island of Jan Mayen
in Greenland Sea; coasts of Newfoundland;
Davis Strait between Canada and Greenland**

STATUS
Estimated population: 500,000 to 600,000

Hooded seal pup after weaning, Gulf of St. Lawrence, Canada. Pups are suckled for 12 days or less and then left by the adults to fend for themselves.

Living on ice

The usual home of the hooded seal is in deep waters on and around the thick, drifting ice of the Arctic Ocean. Toward the end of the summer months the seals are dispersed around the Denmark Strait, and at this time they live solitary lives. Little is known of their winter movements because of the inaccessibility of the northern seas. They are next seen at the time of the harp seal migrations to the Newfoundland area. The hooded seals stay on the outermost fringes of the pack ice, farther from the land than the harp seals. There they congregate in small family groups and soon afterward the pups are born, most births taking place from March onward.

The seals move northward again when they leave the ice around Newfoundland, and they are next seen in the Denmark Strait to the east of Greenland in June and July. Here they gather in groups on the ice, to molt. After the molt is complete, the seals take to the sea again, and from then until they appear off Newfoundland the following year their movements are something of a mystery. They are not seen in great herds. It is thought that they stay feeding in the seas around Greenland.

Mainly fish-eaters

Details of the hooded seal's diet are not very well known, except that they are deep divers and feed principally on larger animals, such as squid and fish, in addition to small prey. Like most seals, they probably feed on whatever happens to be readily available, including other cephalopods, most fish species, shrimps and mussels.

Blue-backed pups

During the month of March the seals congregate on the pack ice. They form family groups consisting of one male, one female and her pup. Females first breed at around 3 years of age. Male seals, on the other hand, may not compete successfully for mates until they are around 13 years, although they reach sexual maturity at 4 to 6 years. Gestation is almost a full year, including delayed egg implantation of 4 months.

At birth the single pup is about 3½ feet (1 m) in length and weighs some 50 pounds (23 kg). It sheds a coat of white, woolly hair just before it is born, much as the common seal does. It emerges with a very short-haired, silvery blue coat on the back, which is sharply distinct from the creamy white color of the underside. The pups with this first silvery coat are known as blue-backs. They are suckled for up to 12 days. If the pup is threatened the adults will defend it. Then the adults mate again and leave the pups to fend for themselves. The pups stay on the ice for a further 2 weeks and then take to the sea.

Threat posed by polar bears

The hooded seal's greatest predator is probably the polar bear, which inhabits the same seas. These bears are more dangerous to the pups than to adults. Humans also hunt the hooded seal for its pelt, with the young blue-backs being the hunters' main target. Sometime hooded seals are hunted by the Inuit, who eat their flesh.

Inflatable nose

When the hood of the male is inflated, it is about 8–9 inches (20–23 cm) high and about 12 inches (30 cm) long. It is red in color and made up mostly of elastic tissue. There is no lining of blubber under the skin, but inside it is lined by an extension of the nasal membranes and it is also divided down the middle by the nasal septum. The hood itself is inflated by air from the stomach and throat. When it is deflated, the hood or sac hangs down over the end of the nose rather like the proboscis of elephant seals.

It is not clear what function the hood serves. Sometimes an angry seal has been seen to inflate its hood, but this is not a regular response to danger. Some seals have also been seen to blow out a red or pink bladder from one nostril, but this again only appears occasionally, both when the seal is irate and when totally undisturbed. This bladder is only produced from one nostril. It is part of the membrane lining the nostril, and it is thought that it is inflated by first inflating the hood. As this collapses, the air forces the nasal sac out of one nostril. Occasionally the hooded seal is referred to as the bladdernose because of this.

A male hooded seal with its hood deflated. It is not known what purpose this inflatable sac serves, given that the male will inflate its hood both at times of stress and when completely undisturbed.

HOOPOE

The hoopoe feeds almost entirely on insects and their larvae and pupae. The large, fan-shaped crest is normally only raised when the bird is excited or alarmed.

The hoopoe is very rarely seen in the British Isles. From there its northernmost range runs across to Lake Baikal, north of the Gobi Desert, and the Amur River in Manchuria, China. South of this line the hoopoe is found over Asia, except Korea and the Indus valley, as far south as Malaysia. In summer hoopoes may be found as far north as Iceland and Spitzbergen, but they do not breed in these regions.

In Africa hoopoes are found over most of the continent, except the Sahara Desert and the rain forests of Central Africa. Birds that are resident in sub-Saharan Africa have more white on the wings and are considered by some to be a separate species. In Madagascar the bird is accepted by most people to be a separate species, the Madagascar hoopoe, *Upupa marginata*.

Need perches

Hoopoes prefer warm, dry climates with open woodland where they can perch on branches and descend to feed on the ground. They also visit orchards and gardens. The hoopoes living in the north migrate south in winter, the European population moving to Africa. When the climate was warmer hoopoes used to breed farther north.

The name of the hoopoe is derived from its *hoop-hoop-hoop* call and has resulted in the scientific name of *Upupa epops*. In the breeding season the call is accompanied by a fluffing out of the neck feathers and bowing.

AT FIRST SIGHT THE HOOPOE is not always the striking bird that is depicted in pictures. Except when the bird is excited, the fan-shaped crest lies folded over its head. Nevertheless, its plumage is very handsome. The hoopoe is pinkish brown in color except for the conspicuous black-and-white bars on its lower back and wings, and its black tail with a white bar at the base. The hallmark of the hoopoe is its crest of chestnut or pink-brown, black-tipped feathers. The bird is about 1 foot (30 cm) long, including its 3–3½-inch (8–9-cm) tail and a slender, curved bill of 2⅓ inches (6 cm) in length.

The hoopoe lives in most of the Old World. It breeds throughout Europe, except in Scandinavia, and ranges as far north as the Gulf of Finland.

Probing for food

Hoopoes feed on the ground, flying to cover with slow, undulating wingbeats. Their long, curved bills are used to probe the soil for insects and their larvae, large grubs such as cockchafer grubs, spiders, worms, large centipedes and occasionally lizards. In the winter they also eat ant lions and termites. The hoopoe frequently probes in animal droppings, and even carrion, for insects.

Nesting in holes

The hoopoe lays its eggs in holes in trees and in buildings, sometimes under piles of stones. It will also nest in nesting boxes. Occasionally it makes a rough structure of grass or feathers, but otherwise the eggs are laid on the floor of the

HOOPOE

CLASS	**Aves**
ORDER	**Coraciiformes**
FAMILY	**Upupidae**
GENUS AND SPECIES	***Upupa epops***

WEIGHT
1⅗–2⅘ oz. (45–80 g)

LENGTH
Head to tail: 6½–8 in. (16.5–20 cm)

DISTINCTIVE FEATURES
Large, pink-brown crest, tipped with black; long, down-curved bill; pinkish brown head, back and underparts; black-and-white barred wings and rump; long black tail

DIET
Almost entirely insects; also centipedes, spiders, worms and small lizards

BREEDING
Age at first breeding: 1 year; breeding season: eggs laid mid-February to May (North Africa), late April to early May (Europe); number of eggs: 7 to 8; incubation period: 15–16 days; fledging period: 26–29 days; breeding interval: 1 year

LIFE SPAN
Not known

HABITAT
Open country with numerous exposed surfaces and perches; nests in old trees

DISTRIBUTION
Summer: throughout Europe south of Gulf of Finland; Central Asia east to eastern China; Nile River region of Egypt. Winter or resident all year: parts of North and sub-Saharan Africa; western Middle East; southern Asia.

STATUS
Common

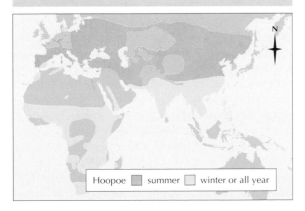

Hoopoe ▮ summer ▯ winter or all year

hole or crevice. The female lays seven to eight light gray to olive eggs. She incubates them by herself for 15 or 16 days. The female never leaves the nest during this period, and all her food is brought by the male. Meanwhile a pile of droppings builds up in the nest hole, staining the eggs. The female hoopoe has a strong odor derived from an oily secretion from the preen gland. Apart from spreading this over her feathers during preening, both the female and the chicks can eject the oil as a musky spray when frightened. At first the female broods the almost naked young while the male brings food for them in his bill. Later both parents feed the chicks, which leave the nest after 26–29 days.

Looking fierce

Hoopoes can move a surprising speeds, although their undulating flight and slow wingbeats give them the appearance of being slow fliers. They can even elude trained falcons flying above them. These birds also have a strange display, apparently used to scare away raptors (birds of prey). On seeing a hawk, a hoopoe spreads its wings and tail, displaying the bold black-and-white stripes. At the same time the head is raised and the bill is pointed into the air.

Wisdom of Solomon

Hoopoes have long been persecuted because of their alleged medicinal properties. They appear repeatedly in folklore. The ancient Egyptians, for example, held the hoopoe in great esteem and its head appears as a hieroglyph. One tradition tells of how hoopoes originally bore crests of gold and that this led to their being killed. The hoopoes petitioned King Solomon, as he could understand the language of animals, to ask for divine help. As a result, the hoopoes were granted crests of feathers instead of gold.

An adult hoopoe in Spain. The name is derived from the bird's highly distinctive **hoop-hoop-hoop** *call.*

HOPPER

THERE ARE SEVERAL THOUSAND species of hoppers belonging to five different families commonly called froghoppers. These families are the froghoppers themselves, and the leafhoppers, treehoppers, planthoppers and jumping plant lice. Hoppers are related to cicadas and aphids. All are small, and most of them are tiny. They are, however, of great economic significance because they often suck sap from plants, causing them to wilt, and many inject viruses into the plants, causing diseases.

Froghoppers and cuckoo spit

There are more than 2,500 species of cercopids, or froghoppers, mainly in the warmer regions of the world. Although most froghoppers are dull, there are several colorful *Cercopis* species in Europe. One species, *Philaenus spumarius,* is usually fairly dull but it has 12 different color forms, or morphs, ranging from melanic (dark) to a pale, creamy brown color.

In northwestern Europe the common cuckoo, *Cuculus canorus,* arrives in April on migration from Africa. Soon after it arrives, blobs of spittle begin to appear on plants. An ancient belief, maintained as a tradition in some rural districts, is that the birds have been spitting, giving the name "cuckoo spit" to these blobs of froth.

Inside each blob of foam is a froghopper nymph, a tiny, pale brown insect with a large head. Sinking its proboscis into the skin of the plant, it sucks sap at such a rate that the liquid

A rhododendron leaf-hopper, Graphocephala coccinea, *resting on a developing seed pod. Leafhoppers can do serious damage to plants, including soft fruits and vegetables.*

quickly passes through its digestive tube and out at the other end. There it mixes with a soapy fluid given out from glands on the underside of the abdomen. The sides of the abdominal segments are extended to curve beneath the body, enclosing a cavity into which the spiracles (breathing holes) open. This chamber opens to the rear through a valve, and the froth or spume is caused by expelled air coming into contact with the fluid passing over the valve. The adult froghopper does not make any froth, but leaps from plant to plant. Its large head and jumping powers strongly resemble those of a frog and earn it the name froghopper or spittlebug.

Leafhopper sharpshooters

The Cicadellidae, or leafhoppers, comprise a huge, worldwide family containing nearly 20,000 species. Most of these measure less than ⅓ inch (8 mm). Even tropical giants are less than 1 inch (2.5 cm). Leafhoppers are serious pests in many parts of the world, damaging a wide variety of cultivated plants, including fruits and other crops. Most of them are powerful jumpers. Some are known as sharpshooters from the way they shoot drops of clear liquid, or honeydew, from the tip of the abdomen. Because the liquid is sweet, ants are attracted to it.

Leafhoppers are also known as dodgers. Instead of leaping to safety when disturbed, a leafhopper will often run around to the other side of a leaf or twig. It then returns to see if all is clear, retreating rapidly again if it is not. Leafhoppers are pests on rice, potatoes, beets, grain, grass and soft fruits. The females use a sharp ovipositor to lay eggs in long rows just under the skin of plants.

The jumping plant lice

Also known as lerps, the 2,000 species of jumping plant lice make up the family Psyllidae. They are small or minute insects, usually 2–3 millimeters long, bearing a strong resemblance to cicadas. Many species keep to one particular tree, especially apple and pear trees, and several are responsible for gall formation.

After passing the winter as eggs laid the previous autumn on leaf scars, the nymphs of jumping plant lice damage fruit blossom and stunt shoots by sucking sap. They give out their honeydew in long, slender, waxy tubes. When these break up, the honeydew spreads over the leaves of the food plants. In Australia the Aboriginal peoples collect and eat the honeydew of the lerps that live on mimosa and eucalyptus.

HOPPERS

PHYLUM **Arthropoda**

CLASS **Insecta**

ORDER **Hemiptera**

SUBORDER **Homoptera**

(1) SUPERFAMILY **Planthoppers,** *Fulgoridea*

(2) FAMILY **Froghoppers,** *Cercopidae;*
leafhoppers, *Cicadellidae;* **jumping plant**
lice, *Psyllidae;* **treehoppers,** *Membracidae*

SPECIES **Approximately 32,000**

ALTERNATIVE NAMES
Froghopper: spittlebug (U.S. only);
jumping plant louse: lerp

LENGTH
Froghoppers and leafhoppers: largest species
up to 1 in. (2.5 cm); most species up to ⅓ in.
(8 mm). Jumping plant lice: usually 2–3 mm.
Planthoppers: largest species up to 4 in.
(10 cm). Treehoppers: up to ⅔ in. (1.7 cm).

DISTINCTIVE FEATURES
Usually very small or minute; wings meet
roofwise over body; piercing mouthparts

DIET
Plant sap, usually from xylem (stem)

BREEDING
Do not go through complete metamorphosis

LIFE SPAN
Varies according to species

HABITAT
Trees and other plants

DISTRIBUTION
Virtually worldwide, especially in Tropics

STATUS
Abundant

Planthoppers

The 5,000 species of planthoppers belong to several families in the superfamily Fulgoridea, but at one time were placed in a single family. They include the lantern flies, Fulgoridae, of which there are about 750 species. Some of these reach 4 inches (10 cm) long. There are also the butterfly bugs, known as Flatidae, a name derived from some of the more colorful tropical species, which resemble butterflies in flight. Finally there are the Dictyopharidae, or false lantern flies. These are mostly green or brown bugs that feed on grasses. One, *Laterinia laterinia*, has a head like a peanut, almost the same size as its body. Its eyes are set far back and the markings on it make it look like a minature crocodile's head. Most planthoppers are not pests.

Hoppers in the trees

The last group is the treehoppers, Membracidae. This is a large, worldwide family containing nearly 2,500 species. About ⅔ inch (1.7 cm) is the maximum size for these insects. Numerous species gather in large aggregations. The distinguishing feature is the considerable extension of the pronotum (the area immediately behind the head), which in many tropical species assumes complex and bizarre shapes. *Cyphonia clavata* is one of several tropical species with pronatal adornments that probably mimic an open-jawed ant. As few predators feed on ants, this could be a form of protection. The wings are transparent, which enables the body to blend into the leaf, enhancing the antlike appearance.

Like the other families of hoppers, treehoppers feed on sap as nymphs and adults and are noted for their jumping ability. Most of them are not present in large enough numbers to be pests. One, the buffalo treehopper, *Ceresa bubalis*, may damage fruit trees through laying eggs in holes in the bark. As nymphs, most treehoppers are inconspicuous, but many, especially those in tropical America, have strange shapes. The external skeleton of the front part of the thorax becomes much enlarged and may also take on a an unusual shape. In the thornbug, *Umbonia crassicornis*, it has the shape of a large rose thorn.

Some hoppers have a close relationship with ants, in which the ants care for the hoppers in order to drink the sweet honeydew that they produce. Shown above are treehoppers (Bolbonota sp.) being tended by harvester ants (Pheidole sp.).

HORNBILL

HORNBILLS ARE SO CALLED because of their huge, often strangely shaped bills, which in some species have a large, helmetlike structure called a casque. This gives the birds an unwieldy, top-heavy appearance. The casque, which is larger in males than in females, is usually very light, being made up of spongelike bone. That of the Malayan helmeted hornbill, *Rhinoplax vigil*, however, is solid bone and has been used in the past by humans in carvings. Another feature of some species is the stiff eyelashes, especially obvious in the ground hornbills, which have red patches of skin on their faces and throats but no casque. The rufous-necked hornbill, *Aceros nipalensis*, of India, also lacks a casque but has blue skin on its face and scarlet skin on its throat.

There are about 45 species of hornbill, typically with large heads, thin necks, broad wings and long tails. Their plumage is usually brown or black with white markings. They vary in size between the turkey-sized Abyssinian ground hornbill, *Bucorvus leadbeateri*, which is over 3½ feet (1.1 m) in total length with a 5-feet (1.5-m) wingspan, and the red-billed dwarf hornbill,

Tockus camurus, which is only 1¼ feet (38 cm) long. The range of the hornbills extends through Africa south of the Sahara Desert, but excludes Madagascar. They are also found across tropical Asia from southern Arabia east to the Solomon Islands and the Philippines.

Legendary birds

The habits of most hornbills are not well known despite their striking appearance. Hornbills feature prominently in folklore, and from early times Europeans have brought back strange tales about them. Folklore and superstition often prevented hornbills from being killed, but nowadays their forest homes are being cut down and they are being deprived of the large trees they need for nesting. Exceptions are the two species of ground hornbills, the Abyssinian and the southern ground hornbill, *Bucorvus cafer*, which live in open savanna country.

One hornbill that has been studied is the silvery-cheeked hornbill, *Bycanistes brevis*, of East Africa. It feeds in flocks, often commuting some distance between feeding grounds and roosts, where 100 or more birds spend the night roosting

An Abyssinian ground hornbill preying on caterpillars, Masai Mara Game Reserve, Kenya. Ground hornbills do not have the distinctive casque of many other species.

MALABAR PIED HORNBILL

CLASS	**Aves**
ORDER	**Coraciiformes**
FAMILY	**Bucerotidae**
GENUS AND SPECIES	***Anthracoceros coronatus***

LENGTH
Head to tail: 26 in. (65 cm)

DISTINCTIVE FEATURES
Very large size; massive yellow bill with black and yellow casque (bony, helmetlike structure). Male: upperparts black, except white trailing edges to wings; white underparts; long, black and white tail. Female: less yellow on bill and casque.

DIET
Fallen fruits; termites and other insects; occasionally fish and nestlings

BREEDING
Age at first breeding: not known; breeding season: March–September; number of eggs: 1 to 3; incubation period: probably 25–40 days; fledging period: probably 45–50 days; breeding interval: not known

LIFE SPAN
Not known

HABITAT
Moist tropical forest

DISTRIBUTION
West coast and parts of central India; throughout Sri Lanka

STATUS
Uncommon, but not under threat

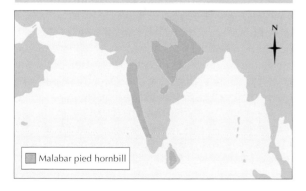

Malabar pied hornbill

together in tall trees. They fly back and forth in small parties making a great deal of noise, including loud cries, roars and bellows. The air rushing through gaps in their wings where the plumage feathers do not cover the bases of the flight feathers makes a loud puffing sound.

Mainly fruit-eaters

Hornbills feed largely on fruits such as figs, passion fruit and various berries, which are collected by hopping about the larger limbs of trees. The larger hornbills can swallow palm nuts, and in Southeast Asia hornbills eat the fruit of *Nux vomica* and related plants that contain strychnine. The hornbills are safe because they do not crack the seeds that contain the strychnine. Some hornbills feed mainly on insects, with termites being a favorite food, and a wide range of larger animals might also be taken, including lizards. Nests are robbed for their eggs and young chicks, and hornbills take advantage of forest fires to catch the animals fleeing the flames. The white-crested hornbill, *Tropicranis albocristatus*, often follows troops of monkeys through the forests of West Africa, feeding on the insects they disturb, while other species follow army ant columns for the same reason.

Hornbills feature prominently in folklore. The rhinoceros hornbill, Buceros rhinoceros, *shown above, is revered in Borneo, where it is often depicted on carved effigies.*

An African red-billed hornbill, Tockus erythrorhynchus. *Hornbills have unusual breeding habits. In most species the female is walled into the nest while rearing her brood of chicks and might remain imprisoned there for up to 4 months.*

Hornbills also attack snakes, sometimes several hornbills banding together to kill one large snake. The hornbills rain blows on the prey with their large, sharp-edged bills, at the same time shielding themselves with their wings. This seems to be an instinctive way of attacking such reptiles, with the birds attacking snakes in this way from a young age.

Walled into the nest

Apart from the ground hornbills, which nest among rocks or in tree stumps, hornbills nest in hollow trees, sometimes using abandoned woodpecker holes. After the eggs have been laid the female walls herself in, blocking up the entrance of the hollow until there is only a narrow slit left. She is then fed by the male, which passes food through the slit. It is this habit that has gained hornbills their place in folklore. Among some African tribes they are considered to be symbols of marital fidelity.

The walling-up behavior varies from species to species. Sometimes the male helps collect material or builds the wall, while in other species the female imprisons herself unaided. The female silvery-cheeked hornbill builds the wall from the inside with a plaster of mud and saliva that is brought by the male, while other species use dung, clay or droppings. While she is still able to, the female squeezes through the gap in the wall, but when it is reduced to a slit, she is completely imprisoned and must rely on the male to bring food. He props himself against the trunk using his tail and regurgitates food for her.

The usual clutch is three to five eggs, but the larger hornbills may lay only one or two. Incubation lasts 1–2 months and the chicks take 6–7 weeks to fledge, perhaps longer in some species. Some females stay in the nest until the young are ready to fly, while others with larger clutches break out and help the male with the feeding. When the time comes for the hornbill chicks to leave the nest, the wall is broken down and the chicks fly out, some with the encouragement of their parents.

Confinement problems

The value of this strange nesting habit is presumably to prevent the nest from being robbed. No doubt it is very successful, as the wall sets rock-hard and any predator trying to get through has to contend with the powerful bill of the female hornbill within. During her imprisonment she molts her feathers rapidly, so for a short time she is without her flight feathers. This does not matter unless the nest is broken open.

To reduce crowding in the nest both the female and young hornbills sit with their tails folded over their backs. It might be thought that nest sanitation would be a more serious problem. However, although a female hornbill may spend 4 months in its nest, it emerges quite clean, as do the chicks. Nest sanitation has not been fully studied but it is known that fruit stones are cast out and that the excreta is ejected forcibly through the entrance slit. Live insects such as bagworms, beetles and cockroaches are found in hornbill nests, and they may help clear up debris.

HORNED FROG

SOME HORNED FROGS ARE actually hornless, and a few are armored or differ in other respects. Nonetheless, all have a reputation for being powerful and pugnacious creatures.

Horned frogs are large amphibians, up to 10 inches (25 cm) in total length. They are unusually broad frogs with large, bulky heads and wide mouths. Their bodies are ornamented with geometric patterns of green and yellow, or rusty red and yellow, on a blackish background color. The horned frogs' skin is covered with warts on the upperparts and is finely granular on the underside. In some species the eyelids are drawn out into what look like small horns. These are, however, only flaps of skin, neither hard nor sharp. Nevertheless, they add to the distinctive appearance of these frogs.

Two separate families

The dozen or so species of horned frogs of the family Leptodactylidae, genus *Ceratophrys*, live mainly in South America, as far south as northern Argentina, including Uruguay and Brazil. There are also several horned frogs of a different family, Pelobatidae, genus *Megophrys*, in Southeast Asia. These range from China south through Indochina and the Malay Peninsula into Indonesia and the Philippines.

Puzzle over colors

Most horned frogs are terrestrial species. They are, in places, abundant near rivers and swamps, and after rain can be seen crawling through the grass in large numbers. At other times they lie buried in soft ground or among leaf litter with only their backs and heads exposed. Some species burrow, while others will form a cocoon to avoid desiccation (drying out). Probably the best known of these frogs, the ornate horned frog, *Ceratophrys ornata*, does this in the drier parts of its range.

In spite of their striking colors, horned frogs are hard to see so long as they remain still. They give an example of how difficult it is sometimes to decide the survival value of an animal's colors. Lifted from their natural surroundings, the horned frogs' colors are conspicuous, yet their discontinuous pattern is one normally associated with camouflage. What is more, some people claim that the bright colors of horned frogs serve as a warning.

Warning coloration in animals is something flaunted to ward off would-be attackers, quite the reverse of camouflage.

Ferocious frogs

Whatever the meaning of their colors, the fact remains that horned frogs, in marked contrast to most other frogs, can be extremely aggressive. It is said that horned frogs in captivity often leap at their owners at feeding time, biting the owners' fingers and hanging on like a bulldog. In fact this is no more than the horned frogs' normal way of feeding. In the wild the adults do not go in search of food but lie half-buried, waiting to ambush any prey items that come near. They then seize their victims, often jumping out from their hiding places to do so. The prey, particularly of the larger, more voracious species such as Bell's horned frog, can be almost anything moving that the frogs are able to swallow. This includes insects, other frogs, lizards and even snakes, small birds and rodents.

Horned frogs are thought to be strongly cannibalistic and this may control their numbers, for they have few other predators. Although they are said not to be venomous, their aggressive actions no doubt help to deter most animals that might otherwise prey on these frogs. In addition, some species, including Wied's horned frog,

An Asian horned frog, Megophrys nasuta, *in the Malaysian rain forest. Asian horned frogs belong to a different family than the South American species. They are superficially alike, but differ in their anatomy.*

A young Colombian horned frog, Cerato-phrys calcarata, *South America. Although nonvenomous, horned frogs have a reputation for being powerful and aggressive creatures.*

HORNED FROGS

CLASS	**Amphibia**
ORDER	**Salientia**
FAMILY	**South American horned frogs, Leptodactylidae; Southeast Asian horned frogs, Pelobatidae**
GENUS AND SPECIES	**South America: ornate horned frog,** *Ceratophrys ornata*; **about 10 others. Southeast Asia: Asian horned frog,** *Megophrys nasuta*; **several others.**

LENGTH
Head and body: up to 10 in. (25 cm)

DISTINCTIVE FEATURES
Large, broad body; big, bulky head; wide mouth; strong jaws; often blackish in color with geometrical patterns of green and yellow or rusty red and yellow; warty skin; fleshy "horns" over eyes (some species only); bony shield covering head and part of back (Southeast Asian species only)

DIET
Adult: insects; also rodents, lizards, snakes, small birds and smaller frogs (larger species only). Tadpole: small prey (South America); algae and other plant matter (Southeast Asia).

BREEDING
Varies with species. All lay eggs in water and have aquatic feeding tadpoles.

LIFE SPAN
Ornate horned frog: up to 13 years

HABITAT
Floor of tropical rain forests

DISTRIBUTION
Ceratophrys: **Central and South America.** *Megophrys*: **from southern China south through Indochina and Malay Peninsula to Indonesia and Philippines.**

STATUS
Not known

Ceratophrys varia, of Brazil, have a dense bony shield covering the head and part of the back. This probably makes them hard to handle or swallow.

More ferocious frogs

In Southeast Asia there are other species known as horned frogs, belonging to a separate family. This means they differ from the South American horned frogs in their anatomy but resemble them in outward appearance. They have a "horn" on each eyelid and have wide mouths and strong jaws like the South American species. These frogs are also strongly cannibalistic and have a bony shield covering the head and part of the back. The horned frogs of South America and Southeast Asia are a good example of convergent evolution in which unrelated animals that have the same way of life are superficially alike.

Two contrasting tadpoles

The two groups of horned frogs differ, however, in the behavior of their tadpoles. Those of the South American horned frogs are predatory from the start, feeding on other small animals. The tadpoles of the Southeast Asia horned frogs, on the other hand, are vegetarians. Some of these have large, funnel-like mouths by which they hang vertically from the surface film of the water. Their mouths are armed with rows of minute, horny teeth, called denticles, which act as rasps to scrape algae and other small growths from the leaves of water plants.

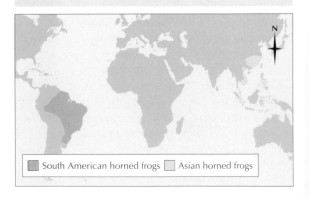

South American horned frogs   Asian horned frogs

HORNED LIZARD

OFTEN REFERRED TO AS horned toads, these reptiles are actually lizards with faces something like that of toads. Apart from their prehistoric appearance, the horned lizards' main distinctive characteristic is their sporadic habit of squirting blood out of their eyes, although this defense mechanism is rarely seen.

Horned lizards measure 1½–4¾ inches (4–12 cm) in total length, depending on species. They have a squat, flattened, almost circular body, short, sturdy legs and a short, slender tail. In most of the species, the head is ornamented with backward directed spines, the so-called horns, and the back is covered with smaller spines. Some horned lizards have been called short-horned, because the head spines are not prominent. Others are called long-horned. In both types the body is covered with small scales, as is usual in lizards, but there are larger, thorn-like scales as well. Usually the edges of the body are ornamented with large, flat scales.

There are 14 species of horned lizards ranging from just over the Canadian border southward through the western United States to Guatemala. The most widely distributed of these is the Texas horned lizard, *Phrynosoma cornutum*, found from Nebraska in the north to Chihuahua and Sonora in Mexico.

Buried in the sand

These lizards are terrestrial and live in arid and semiarid, rocky terrain. They are found from desert and semidesert, sandy country to broadleaf woodland, and from low-lying ground to altitudes of 10,000 feet (3,050 m), depending on the species. They drink dew and hunt insects, particularly ants, but also beetles and grasshoppers.

Horned lizards move slowly forward toward their prey and, when close enough, stop and bend their heads slightly toward it. They then shoot out a thick tongue and in a flash carry the insect back into the mouth. As the day wanes, horned lizards bury themselves in the warm sand. They push their blunt snouts into the sand and by wriggling strenuously make a furrow in which to lie half-buried, or with only the top of the head showing. As autumn approaches, horned lizards spend more and more time buried, and in winter they bury themselves deeper and go into a torpid state.

If horned lizards were larger, they might be mistaken for prehistoric reptiles. Pictured is a short-horned lizard, Phrynosoma douglassi, in Arizona.

Horned lizards are perhaps best known for squirting blood from their eyes as a defense mechanism. This generally occurs only if one of the lizards is attacked by a fox or a dog.

Egg-layers and live-bearers

Most horned lizards are egg-layers. Between April and August the female digs a hole 6 inches (15 cm) deep in the sand. She does this with her forefeet, pushing the sand back with her hind feet. She might lay up to 50 yellowish white, oval, tough-shelled eggs in the hole, although the normal range is 14 to 37 eggs. Each egg is ½ inch (1.2 cm) long. She covers these with sand and leaves. The eggs hatch up to 90 days later, the time varying with the species. A few horned lizards bear live young, up to 30 at a time. These come from eggs that hatch just before being laid. The young measure 1¼ inches (3 cm) at hatching.

A dangerous mouthful

Horned lizards have few enemies because of their armor and the camouflage effect of their colors. Snakes sometimes eat them and often pay for this with their lives, since the horns of the lizard may penetrate the wall of the snake's gullet. They are also well camouflaged, for not only do the lizards bury themselves in the sand, but the mottled colors on their bodies also tend to take on the color or pattern of the sand or gravel on which they are living.

Blood-squirters

Horned lizards are well-known for squirting blood from their eyes when alarmed. This unusual reaction is a defense mechanism, though it is not often seen. Before resorting to such tactics, the lizard's first line of defense is to make itself larger by swallowing air, or it may jab at a predator with its horns. Blood squirting is often a defense against canids such as dogs or foxes. The blood comes from a large sinus behind the eye and emerges from a pore in the eyelid, often squirting for several feet. Although this habit is not unique among animals, it can be particularly spectacular in the horned lizard.

HORNED LIZARDS

CLASS	**Reptilia**
ORDER	**Squamata**
SUBORDER	**Sauria**
FAMILY	**Iguanidae**
GENUS	***Phrynosoma***
SPECIES	**14, including Texas horned lizard,** ***Phrynosoma cornutum***

ALTERNATIVE NAME
Horned toad

LENGTH
1½–4¾ in. (4–12 cm)

DISTINCTIVE FEATURES
Extremely wide, flat body; short, sturdy legs; short, slender tail; head spines and spines projecting from dorsal surfaces of body in most species; small scales covering body, with some larger, thornlike scales; mottled coloration

DIET
Mainly ants; also beetles and grasshoppers

BREEDING
Breeding season: eggs laid April–August; number of eggs: usually 14 to 37; hatching period: up to 90 days; breeding interval: 1 or 2 clutches per year

LIFE SPAN
Not known

HABITAT
Arid and semiarid, rocky terrain; also upland desert and sandy broadleaf woodland

DISTRIBUTION
Extreme southwestern Canada south as far as Guatemala; from western Arkansas to Pacific Coast in U.S.

STATUS
Locally common

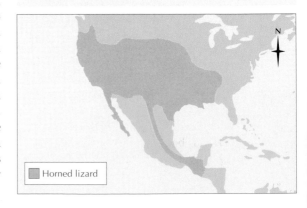

Horned lizard

HORNET

THE NAME HORNET SHOULD really be used only for a large species of social wasp living in Europe, *Vespa crabro*. However, in North America the name is applied to another large native wasp, the white-faced hornet, *Vespula maculata*, as well as to the European species, which has been introduced there. The common social wasps of the United States, genus *Dolichovespula*, popularly known as yellow jackets, and large social wasps in the Tropics are also sometimes called hornets. In the rest of the English-speaking world "hornet" has lost all precise meaning and is applied to any large, social wasp, rather as "tarantula" is often used to describe any large spider.

The European hornet is distinguished from the two almost identical common wasps, *Vespula vulgaris* and *V. germanica*, and other similar species by its larger size and distinctive coloration. It is dull orange and brown instead of the bright yellow and black of these wasps. Worker hornets are rather larger than the queens, which are over 1 inch (2.5 cm) long.

Papier-maché nests

The European hornet resembles the common wasps in that it makes a papery nest of wood chewed to a paste. However, the nest of the hornet is usually built in a hollow tree, occasionally under a bank or in an abandoned building, and for this reason alone often escapes notice. Wasps use hard, sound wood to make their paper, but hornets are less particular and content themselves with soft, decayed wood, sometimes mixing it with sand or soil. The resultant paper is yellowish and rather coarse in texture. The nest is of the usual wasp type, made up of a series of tiers or layers of cells, separated by interspaces, with the cells opening downward. The cells in a hornet's nest are larger but less numerous than in a wasp's nest. The total population of a hornet colony is also smaller.

Not aggressive

Their large size and striking appearance have led to hornets having something of a reputation for ferocity, although this does them a great injustice.

A male European hornet on an apple. Despite their aggressive reputation, hornets are actually less likely to sting than the more abundant wasp species.

The hornet is, in fact, less aggressive than the common wasp and will not sting unless seriously disturbed. It has the unusual habit among wasps of remaining partly active at night. It sometimes goes to the treacle bait that collectors paint on tree trunks to attract moths that fly at night.

Laying the foundation

The life history of the hornet is essentially the same as that of the common wasp. Fertile females, or queens, appear in the nest toward the end of summer. They mate and then find a sheltered place in which to hibernate, while all the other inhabitants of the nest, males and infertile females or workers, die.

In the spring the queens become active again and search for nesting sites, each one founding a separate nest, or colony. The queen begins the paper nest by building a single tier of downward-pointing cells hanging by a stalk from the roof of a cavity. In each cell she lays an egg, and the eggs hatch into larvae. These she feeds until they grow to full size and pupate, each in its own cell. The pupae produce a small brood of workers.

From then on the energies of the queen are devoted to laying more and more eggs in cells built by the workers, which enlarge the nest as their numbers increase. They also forage for themselves as well as for the larvae and the queen. In the late summer queens and males are reared and the cycle begins all over again. As in other social wasps, the larvae are fed by the workers on animal food, such as pulped flies and caterpillars. Often the young are fed insects that are likely to be harmful species in the garden,

so it is not a good idea to kill the queen. The adults live mainly on liquids such as honeydew and other sweet juices from plants. Wasp tongues are not long enough to reach nectar.

Moths mimic hornets

Hornets are imitated by two species of moths in an example of protective mimicry. The hornet moth, *Sesia apiformis,* and the lunar hornet moth, *Sphecia bembeciformis,* are as big as a hornet and have transparent wings and yellow and brown banded bodies. They look so like hornets that few people or birds will touch them. Birds avoid eating wasps and hornets, both because of of their stings and because they have an unpleasant taste.

Male hornet feeding on a sugar solution. Adults will also take liquids such as honeydew. The larvae, meanwhile, are strictly predatory and are fed by workers on pulped insects.

HORNETS

PHYLUM	**Arthropoda**
CLASS	**Insecta**
ORDER	**Hymenoptera**
FAMILY	**Vespidae**

GENUS AND SPECIES   **European hornet, *Vespa crabro*; white-faced hornet, *Vespula maculata*; others**

LENGTH
European hornet. Queen: up to 1⅕ in. (3 cm). Worker: a little larger.

DISTINCTIVE FEATURES
Large wasp. European hornet: dull orange and brown in color, rather than the bright yellow and black of many other wasps.

DIET
Adult: fluids such as honeydew. Larva: pulped insects.

BREEDING
Breeding season: end of summer, first batch of eggs laid in spring. First larvae reared by queen. Toward end of summer males and queens reared in special, large cells.

LIFE SPAN
Queen: 1 year or more

HABITAT
Nest in hollow trees, both upright and fallen

DISTRIBUTION
European hornet: most of Europe except far north; introduced to North America. White-faced hornet: much of North America.

STATUS
Common

HORSEFLY

T HIS NAME IS GIVEN TO certain large flies of the genera *Tabanus* and *Haematopota*. The females feed on the blood of animals, especially horses and cattle, while the males feed on nectar from flowers. The name horsefly is often also used to describe all the flies of the family Tabanidae, all of which feed in the same way.

Tabanids include the flies called clegs, which are particularly fond of human blood and can be a serious nuisance in warm weather on moors and in woodland. The deerflies, genus *Chrysops*, also belong to the Tabanidae. The true horseflies are sometimes called stouts, probably because of their size. One of the largest species, *Tabanus sudeticus*, may be 1 inch (2.5 cm) long and has a wingspan of nearly 2 inches (5 cm).

Iridescent eyes

Horseflies and clegs are dull gray or brown with clear or mottled wings, but the equally blood-thirsty deerflies are more brightly colored insects, brown and yellow, the wings clear with distinct brown markings. In many tabanids the eyes show brilliant iridescent colors, with rainbowlike bands of gold, red and green. These colors fade after death and are not seen in preserved specimens.

Most horseflies are not important as carriers of disease, although deerflies are known to transmit diseases such as anthrax and tularemia.

The other species are nevertheless harmful to cattle, causing disturbance and loss of grazing time, and consequent deterioration in health. In some parts of the world cattle are driven out to graze through the summer night, but are kept in shelter by day to protect them from horseflies.

Bloodthirsty females

Horseflies may sometimes be seen sitting and sunning themselves on tree trunks and fence posts, but are more often seen feeding on cattle or horses, or flying around them. The females of some species will also attack reptiles and amphibians, but few attack birds.

The clegs and other smaller tabanids might be discovered on human clothing or exposed skin, and are often located by the sharp, painful prick that accompanies their bite. The big horseflies produce a loud hum, but clegs and deerflies arrive silently and stealthily make their way to the nearest area of exposed skin. They are not very quick and are easy to swat, but where these flies occur they are numerous.

Tabanid flies have sharp, bladelike mandibles and maxillae and use these to pierce the skin of their victims. The bite of a large horsefly makes a big hole, and when the fly withdraws its mouthparts a drop of blood will ooze out. Female horseflies feed both by licking up these drops and by sucking from the wound. It has been estimated

Most horseflies are not significant carriers of disease, although one group, the deerflies (Chrysops caecutiens, above), does transmit anthrax and tularemia.

HORSEFLIES

PHYLUM	**Arthropoda**
CLASS	**Insecta**
ORDER	**Diptera**
FAMILY	**Tabanidae**
GENUS	**True horseflies, *Tabanus* and *Haematopota*; deerflies, *Chrysops*; others**
SPECIES	**About 3,000**

ALTERNATIVE NAMES
Stout; cleg

LENGTH
Up to 1 in. (2.5 cm)

DISTINCTIVE FEATURES
Stout body; sharp, well-developed mandibles (mouthparts); bulging eyes, often showing bright, iridescent colors; dull gray or brown body (*Tabanus* and *Haematopota*); brighter, brown and yellow body (*Chrysops*)

DIET
Adult female: blood of large mammals such as horses, cattle and humans; also blood of reptiles and amphibians (some species only). Adult male: pollen, nectar and other juices. *Tabanus* larva: insects and other small prey. *Chrysops* larva: decaying vegetable matter.

BREEDING
Little known. All species holometabolous (go through complete metamorphosis).

LIFE SPAN
Adult: several weeks

HABITAT
Usually damp habitats such as grassland, moors and woodland

DISTRIBUTION
Worldwide except very cold regions

STATUS
Abundant

A female horsefly of the genus Tabanus *laying eggs on a rush overhanging a stagnant pond. Horseflies lay their eggs close to water because the larvae develop in water or moist earth.*

that a single grazing horse or cow may lose as much as 3⅓ fluid ounces (100 cu cm) of blood in this way in the course of a summer day.

Males seldom seen

Our knowledge of horseflies' habits is based mainly on females that come to bite. Males are seldom seen and are comparatively rare in entomological collections. They are known to feed on pollen, nectar from flowers and other juices, but are seldom observed doing so.

In the tropical rain forests of Uganda some of the commonest flies of the forest floor are biting horseflies, but for a long time only females were known. Then, as part of a program of entomological research, platforms were put up so scientists could go up and watch near the treetops or forest canopy. In and above this hitherto inaccessible environment were found swarms of male horseflies, together with a few females. Females are thought to fly up to find a mate and then descend again to resume their quest for blood. Another habit of male horseflies is to dart down to take water from lakes and streams in the same way as swifts, swallows and martins do.

Cannibalistic larvae

Although about 3,000 species of horseflies, in the broader sense, are known, we understand little of their life histories. They are holometabolous, meaning that they go through complete metamorphosis. Eggs are laid on plants or stones close to water. The larvae that hatch then live in water or in moist earth and pass through a number of stages, molting between each one. They are elongated creatures, with rather leathery skin and a breathing tube or siphon at the tip of the abdomen. The larvae prey on assorted insects and other small creatures. For example, the larvae of

big *Tabanus* horseflies will eat any other insects, worms, snails, tadpoles and even each other. They catch their prey with a pair of strong, curved, vertically moving jaws. The larvae of the marble-winged deerflies live on decaying vegetable matter. Eventually they pupate and adults emerge by inflating a balloon-like sac, the ptilinum, which forces the end off the puparium. All horsefly larvae tend to seek drier surroundings to pupate. Even aquatic larvae leave the water and pupate in mud or damp soil.

HORSEHAIR WORM

THE ADULTS OF these long, thin worms, which as larvae are parasites of insects, are sometimes found in fresh water. They are 4–40 inches (0.1–1 m) in length, but only 1–3 millimeters in diameter. Horsehair worms were thought in the Middle Ages to be born of horsehairs that had fallen into water.

An alternative name is hair worms, their threadlike form being reflected also in their scientific name Nematomorpha. The name comes from the Greek *nema*, "thread" and *morphe*, "shape." The way they become entangled in masses of between 2 and 20 worms has earned them the names of Gordian worms or Gordioidea. This is a reference to the worms' similarity in appearance to a knot, specifically one tied by Gordius, king of Phrygia in around 330 B.C.E. Legend has it that he declared whoever could undo the knot would be ruler of all Asia.

Different colors and ornaments

The adults of a given species vary considerably in length and in color from light tan to dark brown, sometimes yellowish or almost black. The males of some species end at the back in two lobes, while those of other species are simply rounded at the hind end like the majority of females. The hind end of the female of one species, *Paragordius*, is three-lobed. Females are generally longer than males. Enclosing the unsegmented body is a tough cuticle covered, according to the species, with a variety of tiny furrows, warts, spines and other ornaments.

Horsehair worms used to be classed with Nematoda until fairly recently, but they are in fact not very closely related to any other organisms. There are about 320 known species of horsehair worms worldwide. All but one belong to the order Gordioidea. There is a single marine species, *Nectonema*, which is parasitic in its young stages on crabs and other crustaceans, and which is classified separately.

Damp or watery habitats

Adult horsehair worms live in temperate and tropical regions in all kinds of fresh water. Their habitats include mountain streams, temporary pools, marshes, underground waters and even in dogs' drinking bowls. A few semiterrestrial species occur in damp soil. The marine species is found in coastal environments. The females are sluggish and move little, while the males swim only slowly by undulating the body in a whipping action, contracting the longitudinally arranged muscles. They have no circular muscles to assist them in their swimming.

The adults do not feed, their digestive tract being degenerate and apparently functionless. Usually they have no mouth, but this means little because food is not swallowed at any time during the life cycle. In the parasitic stage food is absorbed through the general body surfaces.

Gordian knots are marriage knots

The entangled "Gordian knots" contain mating pairs. The male coils the hind part of his body around the hind part of the female and deposits sperm near her cloacal opening. These migrate into a sperm receptacle where they are stored for a few days, until the time of egg-laying.

The eggs are laid in long, gelatinous strings, a fraction of an inch to more than 1 inch (2.5 cm) long. These swell on contact with water to produce masses that are often larger than the parent worm. They are often attached to stones or weed, and may contain several million eggs, each 0.05 millimeters in diameter. Like so many

A "Gordian knot" of horsehair worms of the genus Gordius *found under a stone in a drying stream bed. The larval stage of these worms is parasitic on insects.*

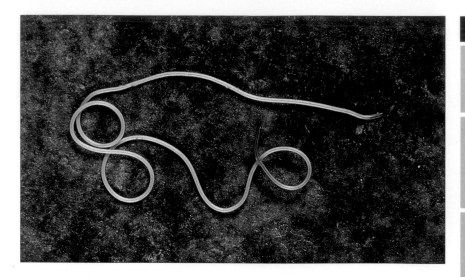

A free-living adult horsehair worm, Gordius villoti. *It is not known exactly how the worm larvae enter the host insect. They may be eaten by the host or they may burrow through its body wall.*

HORSEHAIR WORMS

PHYLUM	**Aschelminthes**
CLASS	**Nematomorpha**

ORDER (1)	**Gordioidea**
GENUS	***Gordius* and *Paragordius***
SPECIES	**About 320**

ORDER (2)	**Nectonematoidea**
GENUS	***Nectonema***
SPECIES	**1**

ALTERNATIVE NAMES
Hair worms; Gordian worms

LENGTH
Total length: 4–40 in. (0.1–1 m); diameter: 1–3 mm; male smaller than female

DISTINCTIVE FEATURES
Long, thin, threadlike worm; usually light tan to dark brown in color, sometimes yellowish or black

DIET
Adult: does not feed.
Larva: parasitic on variety of invertebrates.

BREEDING
Breeding season: spring and early summer; number of eggs: millions; hatching period: 15–80 days for aquatic forms

LIFE SPAN
Not known

HABITAT
Fresh water (most species); damp soil (several semiterrestrial species); coastal waters (*Nectonema* only)

DISTRIBUTION
Worldwide

STATUS
Abundant

parasites, the horsehair worms lay prodigious numbers of eggs to ensure the survival of the species. The males die after mating, and the females die after laying their eggs.

Eaten, or burrowing into hosts?

In aquatic forms the eggs hatch after 15–80 days, depending on the water temperature. The development after hatching is unknown, but it has been suggested that the eggs encyst (develop a protective cuticle) on vegetation and other substrates and are then ingested by hosts feeding on the vegetation. The cyst rapidly degenerates in the digestive tract of the new host. The larva then burrows its way through the intestinal wall into the host's body cavity, continuing its development.

Other researchers have suggested that after the horsehair worm larva emerges from the egg, it will penetrate the body wall of just about any animal it happens to encounter. However, normal development will occur only in specific host animals. After entering the body cavity of an appropriate host, the larva grows to a juvenile stage over a period of several weeks to months, then emerges from the host to mature.

Harmless to mammals

The insects normally parasitized by horsehair worms include a wide range of beetles, grasshoppers, crickets and cockroaches, but these worms have also been seen to emerge from caddis flies and dragonflies. Different species of horsehair worms are thought to vary in the range of insects they parasitize. Many larvae are eaten by other hosts such as the larvae of mayflies and stoneflies, but in these they either die or reencyst in the tissues. If they reencyst the worms may still grow to maturity, provided the host is eaten by a more suitable predatory or omnivorous insect. For example, the giant water beetle, *Dytiscus,* has been known to become infected by eating tadpoles. On rare occasions the larvae turn up in odd places: in a fluke, on the intestinal wall of a fish and in the feces of a child who had perhaps swallowed it in ill-chosen drinking water. Sometimes the presence of horsehair worms in water supplies causes great alarm, and in some regions there is a belief that cows die soon after swallowing such animals. Horsehair worms are, however, almost entirely harmless to mammals.

HORSESHOE BAT

THE HORSESHOE BATS belong to the Microchiroptera, but are readily distinguished from others of this suborder by their nose-leaves. This is the name given to the folds of skin on the faces of many bats. In the horseshoe bats the pattern of the nose-leaves includes one part shaped like a horseshoe. This covers the upper lip and surrounds the nostrils. Above it is a narrow, pointed flap of skin, known as the lancet. Horseshoe bats also differ markedly from other bats in many features of their behavior, especially in the way they use echolocation.

Horseshoe bats' ears are large and do not have a tragus (earlet). Their eyes are small, and their field of vision is obstructed by the large nose-leaf, so sight plays little part in their lives. The females, which are slightly larger than the males, have two dummy teats on the abdomen as well as the functional teats on the chest. The dummy teats are used by the young to hang on to the mother when being carried in flight.

There are about 50 species of horseshoe bats in total, although this number is still under debate. They are found in temperate and tropical regions of the Old World, and all are similar in appearance and behavior.

Varying weight

The greater horseshoe bat, *Rhinolophus ferrumequinum*, of northern Europe, is the species mainly discussed here, but it is typical of the other species in most respects. It has a 2¾-inch (7-cm) head and body with a 1¼-inch (3-cm) tail, and a wingspan of up to 1¼ feet (38 cm). Its weight varies throughout the year from about ½ to 1 ounce (16–30 g), being heaviest in December. Its ears are large, ½ inch (1.3 cm) long, and end in a sharp tip. The greater horseshoe bat has thick woolly fur, ash gray on the back and covering both surfaces of the wing membrane for a short way. The fur of the underside has a yellowish or pinkish tinge.

Late risers

Horseshoe bats tend to come out from their roosts rather late in the evening and often return to roost and fly out again several times during the night. They hunt in dense vegetation and fly low on the wing, sometimes only a few inches off the ground, at most at altitudes of just 10 feet (3 m). The flight of these bats is heavy and butterfly-like, with frequent glides. By day in summer they sleep in caves, tunnels, dark buildings, lofts and roof spaces, and sometimes in hollow trees. They usually roost in colonies, but occasionally singly. Males and females occupy separate roosts.

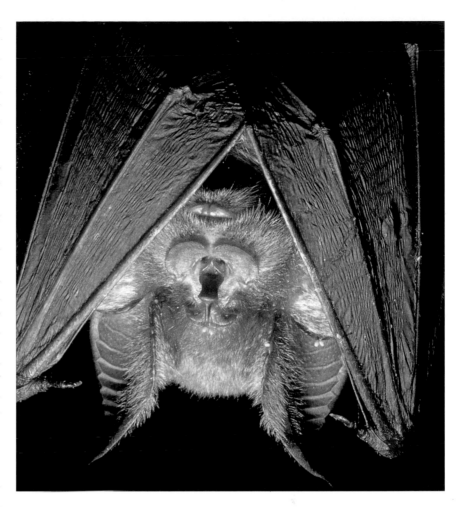

Hibernation

Hibernation in the greater horseshoe bat normally lasts from October to the end of March, although this depends on the weather. Normally the bats sleep singly, in clusters of half dozens or more. Typically they hang by the toes of their hind feet from the ceiling of a cave, with their wings wrapped around their bodies like cloaks.

Hibernation is not continuous, and there is often much movement in the caves. The bats sometimes fly from one cave to another and emerge periodically to feed. The extent to which they move about is influenced by temperature, with bats in one part of the cave moving to another if temperatures fall. Even in a cave, hibernating horseshoe bats sense when the temperature outside rises to 50° F (10° C) or above. They will then come out to feed on dung beetles, which seem to be active during winter, returning if the temperature falls. Further proof that hibernation is intermittent is seen in the accumulations of bat guano (dung) on the floor of the caves during each winter, and the fact that the bats themselves often gain weight during this time.

A greater horseshoe bat at rest. These bats are notable for their nose-leaves, one part of which is shaped like a horseshoe.

Large-eared horseshoe bats, Rhinolophus philippienensis, *roosting in a cave. Horseshoe bats usually roost in colonies, the males and females in separate groups.*

HORSESHOE BATS

CLASS	**Mammalia**
ORDER	**Chiroptera**
SUBORDER	**Microchiroptera**
FAMILY	**Rhinolophidae**
GENUS	***Rhinolophus***

SPECIES **About 50, including greater horseshoe bat, *Rhinolophus ferrumequinum*; and lesser horseshoe bat, *R. hipposideros***

WEIGHT
Greater horseshoe bat: ½–1 oz. (16–30 g)

LENGTH
Greater horseshoe bat. Head and body: 2¾ in. (7 cm); wingspan: 1¼ ft. (38 cm). Other species. Wingspan: 9–22 in. (22–56 cm).

DISTINCTIVE FEATURES
Horseshoe-shaped facial ornamentation; large ears; small eyes; usually brown in color, occasionally red

DIET
Insects, especially moths; also beetles and flies

BREEDING
Age at first breeding: 3 years; breeding season: fall and spring; number of young: usually 1; gestation period: about 42 days

LIFE SPAN
Up to 30 years

HABITAT
Hunting: dense vegetation, usually at low altitudes; roosting: caves and buildings; hibernating: caves

DISTRIBUTION
Southern half of Europe; throughout Africa, Middle East and southern Asia; south through Southeast Asia to parts of Australia

STATUS
Many species common

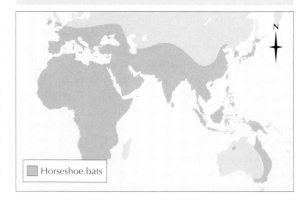

Horseshoe bats

Unusual echolocation

Horseshoe bats use an unusual and advanced form of echolocation to find their way about and to locate prey. Most bats fly with the mouth open, emitting squeaks in regular, short pulses. The echoes from these, bouncing off solid objects, are picked up by the ears, giving the bats a sound-picture of their surroundings. Horseshoe bats fly with the mouth shut, sending out squeaks through the nostrils. They also send out long and powerful pulses at more infrequent intervals, the horseshoe-shaped nose-leaf forming a cup through which the pulses can be beamed. A horseshoe bat, hanging by its feet, can twist itself on its hips through almost a complete circle, scanning the air all around, often darting off suddenly to seize any insect that flies within range.

Horseshoe bats feed on insects, especially moths, but also beetles and flies. They have been seen to settle on the ground to take ground-living beetles. Small insects are devoured immediately, the larger ones being carried to a resting place to be eaten.

Hanging up the baby

Mating is promiscuous, in the fall or in spring. If mating takes place in fall, the sperm lies quiescent in the female's reproductive tract until spring. Then, after a gestation of 6 weeks, the young are born in June and July. The single (sometimes two) young is hairless and blind, and its wings are pale in color. At first the mother takes her young with her when hunting. It clings to her fur with its claws and holds one of the false nipples in its mouth. Later she hangs the baby up in the roost when she goes out foraging. Newborn horseshoe bats have relatively large heads, ears and feet. They mature at 3 years and have been known to live up to 30 years.

HOUSEFLY

MANY DIFFERENT KINDS of flies come into houses. Some are accidental intruders that buzz against the windowpanes trying to get out into the open air again. Others enter houses in the autumn to hibernate in attics and roof spaces. However, there are two kinds that make human habitations their home. One is the housefly, *Musca domestica*, and the other is the lesser housefly, *Fannia canicularis*. The housefly is stoutly built, and in both sexes the abdomen is yellowish or buff in color. Lesser houseflies are smaller and more slender, the females dull grayish and the males similar but with a pair of semitransparent yellow patches at the base of the abdomen. The housefly types are also distinguished by a difference in the veins of the wings, which can easily be seen with a lens. This difference separates the two species, regardless of sex.

Both species have a wide distribution, the housefly being found throughout the Tropics as well as in almost all inhabited temperate regions.

Distinctive flight patterns

Houseflies pass their adult lives in houses, flying about the rooms and crawling over any food that is left exposed. Both species breed in refuse that might accumulate around human dwellings, but their habits differ in detail. Lesser houseflies appear early in the season, while houseflies build up their numbers rather slowly after the winter and are not usually abundant until midsummer.

The males of lesser houseflies fly in a very distinctive way. They choose a spot in a room, often beneath a hanging lamp, and fly as if they were following the sides of a triangle or quadrilateral, hovering and turning sharply at the corners. A single fly will continue to follow the same course for long periods. If, as often happens, more than one fly is patrolling in the same area, one of them will intercept the other and the two whirl together for an instant before parting.

May breed continuously

The breeding habits of the two species are similar, but the larvae of the lesser housefly prefer food rich in nitrogenous compounds, such as urine or bird droppings. These flies are nearly always abundant where chickens are kept. The larvae of the housefly are less particular. Manure and compost heaps and house refuse of any kind provide them with breeding grounds where the larvae will be able to feed.

Like all flies, houseflies are holometabolous, that is, they undergo a complete metamorphosis. The eggs are laid on the larval food, places where

A close-up of the head of a female housefly. Houseflies are able to complete their life cycle from egg to adult in just a week if conditions are warm enough.

the adult flies are also able to feed. The eggs are white and about 1 millimeter long. A female housefly may lay as many as 900 eggs in batches of about 150. The eggs hatch in as little as 8 hours if conditions are very warm, although normally the incubation period is 1 to 3 days.

The white, legless maggots feed rapidly on the decaying material on which they have been laid and undergo several larval stages, molting between each. They may reach full size in under 2 days. They grow more slowly and can live for 8 weeks in colder or less favorable conditions. The pupa is formed in an oval brown capsule called the puparium, which consists of the last larval skin. Instead of being shed at pupation, this is retained and serves the same function as the moth cocoon. At 60° F (15° C) houseflies will breed continuously throughout the year, taking about 3 weeks from egg to adult. In the Tropics the cycle may be completed in just 1 week.

The lesser housefly has a similar life cycle, but its larva is different in appearance, being flattened, with rows of short, branched tentacle-like processes on the upper surface of its body.

Houseflies massed on dung. Adult flies feed mainly on liquid nutrients, their food being partly digested by enzymes in their saliva and then sucked up.

HOUSEFLIES

PHYLUM	**Arthropoda**
CLASS	**Insecta**
ORDER	**Diptera**
SUBORDER	**Cyclorrhapha**
FAMILY	**Muscidae**

GENUS AND SPECIES **Housefly, *Musca domestica*; lesser housefly, *Fannia canicularis***

LENGTH
**Housefly: up to ⅓ in. (0.8 cm).
Lesser housefly: slightly smaller.**

DISTINCTIVE FEATURES
**Housefly: medium-sized and stoutly built; yellowish or buff abdomen.
Lesser housefly: slender build; dull grayish in color; 1 pair of semitransparent yellow patches at base of abdomen (male only).**

DIET
Adult: fluids taken from almost any organic matter. Larva: decaying matter.

BREEDING
Age at first breeding: about 3 weeks (ideal conditions), when slightly older (cooler conditions); breeding season: all year (ideal conditions), during warmest months (cooler climates); number of eggs: about 900, laid in batches of 150; hatching period: 1–3 days, less in warm climates

LIFE SPAN
Several weeks in warm conditions; longer in cooler environments

HABITAT
Usually human habitations such as houses and other artificial structures

DISTRIBUTION
Worldwide

STATUS
Abundant

Flies in winter

Flies disappear in wintertime, and the question of where they go is often asked. There seems no simple answer to it. Houseflies may hibernate as adults or continue breeding slowly in warm places, especially in buildings where cattle are kept. Probably the fly has different adaptations for wintering in different parts of its range. In warm regions it breeds year-round.

Sucking up their food

Adults of both species feed by settling on moist or dry organic matter of almost any kind. Crude sewage and a bowl of sugar are equally attractive, and the insect may fly straight from one to the other. When the fly feeds it extends its proboscis and applies the spongy pad at the end (the labella) to the food surface. Saliva runs out through numerous fine canals known as pseudotracheae, and its enzymes begin to break up the food. The partly digested food is then sucked up.

Flies that have overfilled their stomachs will often regurgitate on any surface on which they happen to be resting, leaving little dirty spots.

People sometimes think that they have been bitten by a housefly. The mistake is excusable because the stable fly, *Stomoxys calcitrans*, looks almost exactly like a housefly. Its mouthparts are, however, very different, consisting of a stiff piercing organ, and it feeds, as horseflies do, by sucking blood. Its bite is quite painful, and it can penetrate skin even through thick clothing. The stable fly breeds in dung mixed with straw and is far less common now than when horses were kept in large numbers.

Bearers of disease

Houseflies have been described as among the world's most dangerous insects. By feeding on excrement and exposed foodstuffs they are potential carriers of gastrointestinal diseases such as dysentery. Houseflies taken from a slum district have been found to carry on average more than 3.5 million bacteria per fly, and more than 1 million were found on flies taken from cleaner districts. These are not all disease bacteria, but some of them are very likely to be. Flies spread germs by regurgitating food that contains bacteria. Also, the flies often defecate while feeding and so many germs are also carried on their feet. Infants and small children often suffer most from fly-borne disease. In a tropical village infant mortality dropped in one year from 22.7 to 11.5 percent when flies were controlled by an insecticide.

Developing an immunity

It is not difficult to kill flies in vast numbers by spraying insecticides on the places where they feed and breed, but they have a remarkable capacity for developing resistance to specific poisons. No individual fly develops resistance during its lifetime, but some will almost always survive a spraying, and these will include individuals having, by an accident of nature, some degree of immunity to the pesticide being used. This immunity is inherited by their offspring in varying degrees, and the most resistant of these will again survive and breed. Selection of this kind continues with every generation until the insecticide is useless in any safe concentration. Such acquired resistance in insects is, in fact, an example of very rapid evolutionary change. It is also one of the most compelling arguments against relying too much on pesticides to control harmful insects.

Improved methods of control

One of the best ways to reduce housefly populations is actually to cover their breeding sites with soil. Control of houseflies can also be achieved by depriving them of breeding places altogether. Modern, more hygienic ways of life have already gone a long way toward doing this. Waterborne sanitation, the use of covered dustbins and the decline of the horse as a means of transportation are three obvious factors. However, flies will be with us for a long time yet, especially in regions with hot climates. The best strategy for controlling them is to exclude them from our houses and, above all, keep them away from our food.

A housefly resting on food. It has been said that the housefly is one of the world's most dangerous insects because of its capacity to spread disease.

HOUSE MOUSE

House mice will eat anything with nutritional value. This includes not only food found in cupboards or the garbage, but also soap and toothpaste.

Stowaways

It used to be thought that the house mouse, like the brown and ship rats, originated in Central Asia. The present view is that its original range probably included the Mediterranean area, both southern Europe and North Africa, and most of the steppe zone of Asia as far east as Japan. Certainly it was known in Europe at the time of ancient Greece. Nowadays it has spread all over the world, largely through being accidentally carried by humans. It is now found wherever there are human habitations, in houses, on farms and in warehouses, in the Tropics as well as in the Arctic. It used to be that mice were taken from country to country across the seas on ships. Modern day stowaways now travel by air as well.

Swift and silent mover

Mainly but not wholly nocturnal, the house mouse moves quickly and silently. It is extremely agile and can climb well up walls of brick or concrete. When suddenly alarmed, it can leap across long distances, especially over vertical barriers. It is also able to squeeze through holes as small as ⅜ inch (9 mm) in diameter. House mice can swim quite well if they have to.

Throughout their present range house mice have become mainly associated with human beings, but not merely in towns. They are also found in isolated buildings, well away from towns and villages. House mice usually live in buildings in warmer countries because they are killed off by predators when in open country. Still, they do sometimes inhabit woods and fields, particularly in the summer months. On some islands where predators are virtually nonexistent, mice have become well established in the countryside. In such situations, with no or very few competitors to keep their numbers down, house mice sometimes become feral, as has happened in New Zealand.

Wild origins

Naturalists believe that there were originally four wild subspecies of house mice. One of these, the original outdoor form, is small with its tail considerably shorter than its head and body length. The other three wild subspecies became

THIS IS PROBABLY THE most familiar and the most widely distributed rodent. As its name suggests, the house mouse lives mainly in or around human habitations and buildings, especially where food is stored. Mice are very adaptable to different environments. For example, they can even be found living in large meat refrigerators, in constant darkness and with temperatures around the freezing point.

The house mouse is small, having a 2½–3¾-inch (6.5–9.5-cm) head and body length. It weighs ⅖–1 ounces (12–30 g) and has a scale-ringed, sparsely haired tail about the same length as its body. Its muzzle is pointed and it has moderately large ears and eyes, although smaller than those of wood mice (genus *Apodemus*). The fur is brownish gray in color, slightly paler on the underparts.

HOUSE MOUSE

CLASS	**Mammalia**
ORDER	**Rodentia**
FAMILY	**Muridae**
GENUS AND SPECIES	***Mus musculus***

WEIGHT
⅖–1 oz. (12–30 g)

LENGTH
Head and body: 2½–3¾ in. (6.5–9.5 cm);
tail: 2⅓–4 in. (6–10 cm)

DISTINCTIVE FEATURES
Small size; smaller eyes and ears than
wood mice; pointed muzzle; gray above;
paler underparts; scale-ringed, sparsely
haired tail

DIET
Originally seeds, grain and insects; adapted
to eat anything with nutritional value, such
as paper, toothpaste and soap

BREEDING
Age at first breeding: 35–45 days; breeding
season: all year; number of young: usually
5 or 6; gestation period: 19–20 days;
breeding interval: 5 or more litters per
year, depending on habitat

LIFE SPAN
Up to 3 years in captivity; considerably
less in wild

HABITAT
Lives alongside humans in all locations;
feral in certain habitats that have very
few or no competitors, for example in
New Zealand

DISTRIBUTION
Worldwide: wherever humans live

STATUS
Generally abundant

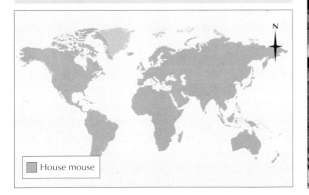

House mouse

commensal with human beings, meaning that they adapted to living with humans, later coming to depend on them for food and shelter. These three have since given rise to numerous other subspecies. Even these have become thoroughly mixed by interbreeding, with a resulting wide range of color varieties, dark and light as well as the typical mouse gray.

Sometimes these indoor mice will live outdoors but, as a rule, this occurs only where there are cultivated crops. There are also the strains of tame and laboratory mice and certain breeds that are hairless except for the whiskers.

Strong smell

House mice have territories that they mark with their urine, and this is responsible for the strong smell associated with mice kept in captivity. It has been found that the more mice cages are cleaned out, the more energetically the occupants mark their cages with urine. As a result, the smell becomes even stronger.

Individual mice occupying a territory can recognize one another by their odor. Experiments have shown that a "foreign" mouse introduced into a territory after having been artificially supplied with the odor of those living in the territory is readily accepted by them. Otherwise such a mouse would be driven away.

If threatened, the aggressive attitude of the house mouse is to rear up on its hind legs and hold its forepaws together, with its nose in the air. Such an attitude can occasionally be seen when a cat is playing with a mouse. The cat drops the mouse, which turns to face the cat as if to fight.

A female house mouse with her young. Litters range from 3 to 13 babies, although 5 or 6 is the usual number.

A female house mouse carrying newborn offspring. The young are born blind and naked but develop quickly and are able to breed after 6 weeks.

domestic residences house mice take food from cupboards and garbage, but they will also eat such unlikely substances as the leaves of house plants, paper, card, electrical cable, toothpaste and even soap.

Those populations that live permanently in meat lockers eat meat alone. Such mice are larger and heavier than usual, have longer coats and make their nests in the animal carcasses. During World War II "buffer depots" of flour were set up in Britain. Mice invading these were able to live only on flour and with very little water. However, house mice that have lived on household wastes and in larders seem unable to revert to natural foods. This was shown by the mice of St. Kilda, an island off the Outer Hebridies, northwest Scotland. When the human inhabitants left the island in 1930, the mouse population soon died out.

Myopic mice

Smell and hearing are the two most important senses for house mice. In the past it was commonplace for those writing about small mammals generally, and mice in particular, to speak of their bright beady eyes, implying keen sight. In fact, mice are myopic and it is doubtful whether in daylight they see much beyond a range of 2 inches (5 cm). How their eyes serve them at night has not been fully studied.

One relatively recent discovery is that house mice and many other small rodents use ultrasonics, sounds that have a frequency above the human ear's audibility limit. Some of their squeaks can be heard by the human ear but others are too high-pitched. A mouse's ears are sensitive to these high-pitched sounds.

Dominant males

The social structure of a mouse colony is loose until the population becomes overcrowded. Then more of a social hierarchy is formed, with one male dominant over the rest; he alone mates with the females. This provides a natural brake to further overcrowding. Nevertheless, there are frequent instances, notably in California and Australia, of mouse plagues, when the ground seems to be alive with tens of thousands of them in quite small areas.

Eat anything edible

House mice mainly feed on seed, grain and insects, but they readily adapt to a wide variety of other foods. In fact they will eat almost anything with nutritional value. In and around

Five litters a year

The success of house mice owes as much to their breeding rate as to their adaptability in feeding habits and finding shelter in buildings. They breed through most of the year. In houses these mice average just over five litters a year, with an average of five or six young to a litter, although the range can be 3 to 12 babies. In cold stores the averages are six litters a year and six young to a litter. In grain stores they average between eight and 10 litters a year. Gestation is 19–20 days, and the young are born blind and naked. They are weaned at 18 days, and reach sexual maturity by the age of about 6 weeks, when they begin to breed. House mice live around 3 years in captivity, but considerably less in the wild.

Driven out of home

Mice living outdoors have many predators. These include carnivores such as weasels, ermines, foxes and cats, a wide range of owls and numerous raptors (birds of prey), specially hawks. Some omnivores will occasionally prey upon house mice, such as crows and rats, all of which normally take small animals. Where house mice abound, snakes often make them their staple diet.

There is a rooted belief in some quarters that rats and mice cannot live together, that if you have mice in a house there will be no rats. It is more likely that when rats take up residence they kill off the mice. Where close studies have been made, it has been found that the house mouse cannot compete successfully even with wood mice (discussed elsewhere), which, although about the same size, are more active.

HOUSE SPARROW

THE NAME SPARROW is given to a number of birds, including the North American song sparrow, *Melospiza melodia*, and its relatives. Old World sparrows (true sparrows) belong to the genus *Passer*, from which the vast order of perching birds, Passeriformes, gets its name. Closely related to the Old World sparrows are the rock sparrows and snow-finches. Together these three groups make up the subfamily Passeridae, which is related to the true weavers such as the bishop birds, fodis and queleas.

Wide-ranging sparrows

The best-known sparrow is the house sparrow, *Passer domesticus*, which has become completely dependent on humans, and has often been deliberately introduced to many parts of the world. About 5¾ inches (14.5 cm) from head to tail, the male has brown upperparts with black streaks, grayish white underparts, a black bib, thin white bars on the wings and a gray rump that can be seen in flight. The bib is more extensive when the male is in breeding plumage. The female has duller plumage, but lacks the male's black bib, the wing bar and the gray rump.

Some closely related sparrows are often confused with the house sparrow. The Spanish sparrow, *Passer hispaniolensis*, of Spain, Italy, Greece and North Africa, can be distinguished by its generally darker plumage, with the black bib extending onto the breast and sides and a brown rather than a gray crown. The tree sparrow, *P. montanus*, has a dark brown crown and a black spot on the whitish cheeks. It breeds in Europe and Asia, from the British Isles to Japan. Other sparrows include the desert sparrow, *P. simplex*, which lives in the deserts of Africa and Asia. The rock sparrows or petronias live mainly in Africa, with three species extending into Asia and one into Europe. The snow-finches are found in mountains from the Pyrenees to Mongolia.

Community life

House sparrows live in flocks with strong bonds among the members. They feed, dust-bathe and roost together, and when one gives the alarm they fly to cover in a tight group. The members of a flock communicate by means of simple calls, the familiar chirruping and cheeping. The call is basically a means of identification and keeps the

Male house sparrow feeding young, England. After being spread and deliberately introduced from Europe, the house sparrow now ranges across much of the world, including North and South America, South Africa, Australia and New Zealand.

House sparrows mainly feed on grain and can strip a crop in a very short time. Their dry diet means that they are thirsty birds.

HOUSE SPARROW

CLASS	**Aves**
ORDER	**Passeriformes**
FAMILY	**Ploceidae**
GENUS AND SPECIES	***Passer domesticus***

LENGTH
Head to tail: about 5¾ in. (14.5 cm); wingspan: 8–10 in (21–25.5 cm)

DISTINCTIVE FEATURES
Strong, conical bill; gray crown; grayish white underparts; chestnut brown and gray upperparts with black streaks. Breeding male: thin, white wingbar; gray rump; extensive black bib. Nonbreeding male: small bib. Female: no bib; browner overall.

DIET
Mainly seeds, particularly grain; also shoots, berries, insects and household scraps

BREEDING
Age at first breeding: usually 1 year; breeding season: varies widely according to latitude and hemisphere; number of eggs: usually 3 to 5; incubation period: 11–14 days; fledging period: 11–19 days; breeding interval: 3 or 4 broods per year

LIFE SPAN
Up to 13 years, usually much less

HABITAT
Generally near human habitation, including villages, suburbs and city centers; often in cornfields and other crops in fall and winter

DISTRIBUTION
Europe, Middle East and Asia, except northern Siberia and parts of Central Asia and China; North and Central America, except far north; also in parts of South America, Africa, eastern Australia and New Zealand

STATUS
Common or abundant in much of range

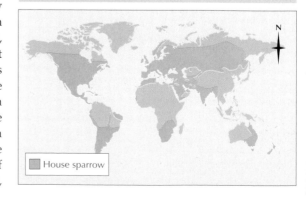

House sparrow

flock together as the birds go about their activities. This communication is particularly noticeable when a flock gathers in a tree before flying to its roost. The places where these chattering masses gathered in London, England, were once called "sparrows' chapels." The call is also used by the male sparrow as a song to proclaim the ownership of his nest site and to attract a mate, while variations are used to give alarm.

Mainly grain-eaters

Sparrows are seed-eaters, the house sparrow feeding mainly on grain. When they occur in large numbers, sparrows often become pests, because a flock can strip a seed crop and beat down the stems in a very short time. Numerous campaigns have been mounted against house sparrows, the most spectacular being that in China, which involved the mobilization of the human population. There has, however, been little success in eliminating these birds. They are very adaptable, being able to feed on all sorts of food such as kitchen scraps and other refuse, while in the days of horsedrawn traffic, urban house sparrows lived largely on spillings from nosebags and undigested grains from droppings.

The natural food of house sparrows includes insects, worms, buds and fruit. They also gather food in the same ways as several other birds. They fly out after insects from perches in the same way as flycatchers, hover and pounce like kestrels, search leaves like tits and flutter after low-flying insects like wagtails.

Wedding parties

The Old World sparrows build domed nests of grass, usually in a hole in a tree, in crevices in buildings or in the nests of other birds such as swallows and martins. The Spanish sparrow nests in colonies, with several hundreds of nests all built in a few trees. The chestnut sparrow, *Passer eminebey*, meanwhile, nests in communities, taking over the nests in the huge communal structures built by the social weavers.

House sparrows are found near their nest holes for most of the year, and in winter they may roost in them. At the start of the breeding season the male advertises his nest site by chirruping, and females are attracted to inspect the site. There is little ceremony in courtship but there is an activity called the "sparrows' wedding" or "sparrows' party," in which a twittering party of males chases a female, which may turn on them and attack. Mating does not take place, and the significance of this activity is not known.

Both sexes build the nest and line it with feathers, and both incubate the clutch of whitish brown-mottled eggs. The clutch usually numbers two to four eggs in the Tropics and three to five in cooler parts. Incubation takes 11–14 days. The chicks leave the nest after some 2 weeks and there may be three or four broods in one year.

Constant companions

The house sparrow is unusual among birds in that it often lives in artificial habitats. It is presumed that it evolved from the Spanish sparrow because, although the two species are separate in Europe, they interbreed in North Africa. The tree sparrow also lives with humans in many parts of the world, but where the house sparrow is spreading, such as in India, it is driving the tree sparrow out of the towns.

Part of the house sparrow's success is probably due to its wide range of feeding habits and its ability to nest on houses. This allows it to nest in the center of cities, and in big warehouses it is possible that the house sparrows pass their whole lives without venturing outside. A pair once raised a brood down a coal mine in Yorkshire, England, though the young did not survive long.

Being so adaptable, the house sparrow has followed Europeans as they spread to new continents, and it is now abundant in North and South America, South Africa, Australia and New Zealand. It has been deliberately introduced into some places and in the past has even been protected by law. Nonetheless, despite its adaptability, the house sparrow's populations have fallen markedly in Britain in the last 20 years. This has been attributed to reduced opportunities to nest in "modern-built" houses and less autumn stubble available to feed on.

A male house sparrow watches over his territory from the top of a pan-tile roof. House sparrows are unusual in that they often live in artificial habitats such as cities.

HOVERFLY

little rocking movements while accurately maintaining their position. This is a remarkable feat of controlled flight, for the air in which the fly is poised is not motionless. There is almost always some lateral drift, as well as eddies caused by rising air currents and the breeze passing around branches. The insect must therefore constantly make minute adjustments of its wingbeats to avoid being carried up and down.

Developed sense of sight
It seems most likely that the hoverfly maintains its position through its sense of sight. Insects' eyes are less efficient than ours at forming images, but much more efficient at detecting small movements. The slightest shift in the fly's position is thus instantly perceived and as quickly corrected, so it remains motionless, not in relation to the air around it, but in relation to the solid objects within its field of vision.

The eyes of hoverflies are relatively enormous. Other insects with very large eyes, such as the dragonflies and robber flies, need efficient vision to hunt winged prey in the air, but hoverflies feed from flowers and the need for accurate control of their hovering provides the only explanation of their very highly developed sense of sight.

Why do they hover? In a few cases it seems to have some connection with courtship and mating, but in most species both sexes constantly hover without taking any notice of each other at all. Another curious habit hoverflies have is of continuing to buzz after they have settled and ceased to move their wings. The sound seems to be produced by vibration of the thorax, but why they do this is not known.

Many kinds of larvae
Most hoverflies have a short proboscis with a spongelike expanded end, which they use to mop up sugary liquids. They can feed only on flowers in which the nectaries are exposed, such as those of ivy. Others have a kind of snout that they can push into bell-shaped flowers, and one East Asian genus, *Lucastris*, has a long proboscis and can feed from tubular flowers. As well as taking nectar, hoverflies also take the honeydew of aphids from leaves.

The feeding habits of the adult hoverflies are fairly uniform, but those of the larvae are extremely diverse. Larvae may hunt aphids, feed on decaying organic matter, feed on the juices oozing from wounds in trees, burrow into stems and roots of living plants or feed on waste matter in the nests of bees, wasps and ants.

A migrant hoverfly, Syrphus balteus, visiting willow herb. Hoverflies are remarkable for their skill in flight, being able to hover virtually motionless even in rising air currents and strong breezes.

HOVERFLIES ARE PROBABLY the most skillful of all insects in flight. They are two-winged flies belonging to the family Syrphidae. Many visit flowers in large numbers to feed on nectar and they are second only to bees in importance as flower pollinators. In North America they are also called flower flies.

Most hoverflies have a superficial resemblance to wasps and bees, being either marked with black and bright colors in contrasting patterns or covered with a coat of short, dense hairs, also variously patterned. In some cases there is such close resemblance between certain species of hoverflies and the wasps and bees living in the same area that there seems no doubt that mimicry is involved. Hoverflies are almost all harmless, and many are useful.

Living helicopters
Hoverflies are most active in sunshine and warm weather. They can be seen in large numbers hovering over flowers with exposed nectaries (plant glands that secrete nectar), feeding from them. When hovering, they sometimes make

The aphid killers include some of the most abundant and familiar hoverflies. Their larvae are sluglike, with the body tapering to a kind of neck at the front end. They have no eyes and the head is small and no broader than the neck. They hunt the swarming aphids by touch, crawling among them and swinging the trunklike neck from side to side. Starting, when very tiny, with three or four aphids a day, a hoverfly larva may be eating 50 or 60 a day when fully grown.

HOVERFLIES

PHYLUM	**Arthropoda**
CLASS	**Insecta**
ORDER	**Diptera**
FAMILY	**Syrphidae**
GENUS	***Eristalis, Syrphus, Episyrphus, Volucella* and *Chrysotoxum*; others**
SPECIES	**Many**

ALTERNATIVE NAMES
Flower fly; drone fly (*Eristalis* species only)

LENGTH
About ⅕–⅗ in. (0.5–1.5 cm)

DISTINCTIVE FEATURES
Adult: very large eyes; 1 pair of wings; many species resemble wasps or bees, with contrasting patterns of black and bright colors; some species covered in short hairs. Larvae: often colorful in predatory species.

DIET
Adult: nectar, honeydew and pollen. Larva: various foods, including sap, stems and roots of trees and other plants, rotting vegetation, aphids (predatory species only), dung and carrion.

BREEDING
Undergo full metamorphosis. Larva has several stages, molting at end of each; 1 to 3 generations per year.

LIFE SPAN
Adult: several weeks. Larva: from 10 days to about 1 year.

HABITAT
Woodland, grassland and areas near water

DISTRIBUTION
Almost worldwide

STATUS
Most species common

One of the most interesting of the larvae living on decaying matter is that of the drone fly, *Eristalis tenax*. Several *Eristalis* species are given this name because they resemble the drones of honeybees. The larvae, often called "rat-tailed maggots," live in the puddles that collect around manure heaps, in water containing little oxygen. At the hind end of its body the rat-tailed maggot has a breathing tube or siphon, which is extensible like a telescope. Its length can be adjusted to reach 4 inches (10 cm) to the surface if need be.

Hoverflies that burrow into living plants provide exceptions to the rule that hoverflies are harmless. The narcissus fly, *Merodon equestris*, spends its larval stage inside narcissus and daffodil bulbs, eating and destroying them. In places where bulbs are cultivated on a large scale, these larvae may cause considerable losses. The adult fly is large and looks like a bumblebee.

Feeding on bees' litter

The big hoverflies of the genus *Volucella* provide the larvae that feed on waste in the nests of bees, wasps and ants. The females enter the nests and lay their eggs, and the larvae from them live on the bodies of dead bees and dead larvae and on any other edible debris. In all cases the egg-laying females and the larvae are accepted by the bees and wasps, which are generally most intolerant of trespassers. In the case of the common species, *Volucella bombylans*, which usually lays in the nests of bumblebees, the adult flies closely resemble the bees. Furthermore, this species occurs in two distinct forms, each of which looks like a particular species of bumblebee. It is tempting to think this helps them get into the nests, but this is far from certain since they also breed in the nests of wasps, which they do not at all resemble.

Many hoverflies (Chrysotoxum cautum, below) bear a close resemblance to bees or wasps. It is not always clear what purpose this apparent mimicry serves.

HOWLER MONKEY

A young black howler monkey in Belize. Howler monkeys have few predators, but some subspecies are under threat as a result of hunting and habitat loss.

THESE, THE LARGEST OF the South American monkeys, are named for their loud calls. Howler monkeys are also called howlers. They have an enormously enlarged hyoid bone in the throat, which forms a shell-like resonating chamber (corniculum). This produces the characteristic loud voice. The hyoid is smaller in the female than in the male but is still quite effective.

A howler monkey's whole skeleton and cartilage are adapted to produce its loud, distinctive vocalizations. For example, the shape of its skull and throat is modified by the vocal apparatus. In connection with the enlarged hyoid, the lower jaw is expanded at its angle, and the skull is long and low, giving the howler a sloping face. The expanded hyoid bone also makes the neck thick and heavy. The male howler monkey has shaggy hair around his neck, making him look even larger.

Largest South American monkeys

Howler monkeys are stoutly built and have a hunched appearance. They are usually black, brown or red in color, but this differs among the nine species. Species include the mantled howler monkey (*Alouatta palliatta*), red howler monkey (*A. seniculus*), brown howler monkey (*A. fusca*) and black howler monkey (*A. caraya*). Their tails are thickly furred and prehensile, with a naked area on the underside at the tip. Howlers have large hands and feet, but like other South American monkeys, the thumb cannot be opposed to the fingers, so howler monkeys pick up objects with the second and third fingers. Both sexes reach a total length of 4–6 feet (1.2–1.8 m), about half of which is tail. Partly because of the modified neck region, the male is heavier, 16–22 pounds (7.2–10 kg), in comparison with the 9–18 pounds (4–8 kg) of the female.

Howler monkeys are widely distributed throughout Central and South America, their range extending north into southern Mexico and south into northern Argentina. They are found wherever there are tropical and subtropical forests.

Howling matches

When howler monkeys wake around sunrise, most groups begin their characteristic howling. The cry is low and resonant in males and a "ter-rierlike" bark in females. One group howling stimulates others to do so, and troops usually howl when they catch sight of each other on the edges of the territory or in the no-man's-land between, warning each other against trespassing. These howling matches maintain the boundaries between neighboring clans, with the cries carrying over a distance of 2–3 miles (3–5 km).

Life at a leisurely pace

The howler monkey's day is taken at a fairly leisurely pace. After howling, the group begins to feed, moving around slowly and often stopping to rest in the sleeping-tree. Then, in the middle of the morning, the animals begin to move out to food trees away from the center of their territory. They then rest again until mid-afternoon. Howler monkeys spend up to three-quarters of their entire day resting. Later they begin to feed again, traveling around and finally calling again before settling down for the night.

HOWLER MONKEYS

CLASS	**Mammalia**
ORDER	**Primates**
FAMILY	**Cebidae**
GENUS	***Alouatta***

SPECIES   **9, including mantled howler monkey, *A. palliatta*; red howler monkey, *A. seniculus*; red-handed howler monkey, *A. belzebul*; black howler monkey, *A. caraya*; and brown howler monkey, *A. fusca***

WEIGHT
9–22 lb. (4–10 kg)

LENGTH
Head and body: 1⅔–3 ft. (54–90 cm); tail: 2–3 ft. (60–90 cm)

DISTINCTIVE FEATURES
Large, strongly built monkey with long tail; thick, heavy neck with (adult male only) shaggy hair; most species predominantly black; others reddish brown with pale face

DIET
Mainly leaves; also fruits (especially wild figs), buds, flowers and nuts

BREEDING
Age at first breeding: 5 years (male), 3–4 years (female); breeding season: all year; gestation period: 180–210 days; number of young: usually 1; breeding interval: 1 year

LIFE SPAN
Up to 20 years

HABITAT
Tropical rain forest

DISTRIBUTION
Central and South America

STATUS
Often locally common but many subspecies declining; critically endangered: *A. fusca fusca* and *A. belzebul ululata*

Howler monkeys

Dominant males

There may be up to 30 howlers in a troop, with normally two to three times as many females as males. In situations where the overall population is thinly spread over a large area, the troops will generally be smaller, each containing one adult male and just a few females and their young. In normal circumstances, however, a troop averages about 18 animals, with several males associating together. Each troop occupies a territory that varies in size according to the amount of food available. As a result, territories change in size and position over a period of weeks. The troop occupies some areas of the territory more often than others and has favorite sleeping-trees.

Within the group, the males are dominant over the females, but not aggressively so. There is a fairly well-defined dominance hierarchy or pecking order among the males, and a less marked and separate hierarchy among the females. The males act in concert in leading the troop and in howling. When the troop meets a neighboring clan, the monkeys act together, the males roaring, females whining, and the young ceasing to play. Howlers are less active on rainy days, when very little roaring can be heard.

Gripping tails

When moving through the trees, the adult males lead the way in order of rank. The whole troop follows, keeping to exactly the same pathway

Howler monkeys move about slowly to feed, taking frequent breaks. Their diet is mainly made up of leaves, but they also eat fruits, buds, flowers and nuts.

A male red howler monkey roaring at dawn to announce his troop's claim to their territory. The modified skeletons and cartilage of howlers enable them to produce their vocalizations.

the tip. This not only provides grip, but is also sensitive to touch. Howlers can use their tails to investigate objects before grasping them.

Feed on leaves

Howler monkeys are vegetarian, feeding on leaves, which comprise about half their daily intake of food. They are adapted for this diet with enlarged sections of their gut containing bacteria that can steadily break down and release the energy in cellulose. Howlers also eat buds, flowers and fruits, along with nuts, including the shells.

They are particularly fond of wild figs, and this is thought to be one of the sources of competition between groups. Each troop needs to have a territory large enough to contain a sufficient number of fig trees. The fruits are pulled in by hand and eaten directly from the stems, but the females sometimes pick and hold food for the young. A great deal of the food, perhaps as much as half, is dropped uneaten, and no effort is made to retrieve it. The prehensile tail is wrapped around a branch for stability while feeding.

Year-round breeding

Breeding takes place year-round in howler monkeys, with just one baby being born after a gestation period of 6–7 months. At any one time, approximately one-third of the females in the troop have infants. The young cling to their mothers' bellies, often with the tail wrapped around the base of the mother's tail. If the infant becomes separated from its mother, it makes a little cry of three notes, while the mother makes a wail ending with a grunt or groan. The infants purr when happy. The young howler monkey is dependent on its mother for about 6 months. Lactation lasts for a year, when a second baby is born, and the first is rapidly displaced. The young play very little and are often as slow and ponderous as the adults.

already taken by the males. Howlers are slow and deliberate in their movements as they travel through the branches. Occasionally they will move more quickly when playing and when excited. They never jump from branch to branch, preferring to form a "bridge," clinging with their prehensile tails and grasping the neighboring branch with their hands. Often an adult will use its body as a bridge in this way to help a youngster move from branch to branch.

Howler monkeys are able to use their long, muscular tails almost as if they were an extra hand. The monkeys are even able to hang from branches by their tails, leaving their arms and legs free. The underside of the tail is bare near

Threats to howler monkeys

Ocelots and other small cats occasionally take young howlers, but there are no serious predators on adult howler monkeys. They suffer to some extent from hunting, but their biggest threat is the clearing of forests in such countries as Brazil. Certain species and subspecies are rapidly declining as a result of such habitat loss.

HUMMINGBIRD

T HERE ARE OVER 320 SPECIES of these minute, brightly colored birds living in the Americas. The largest is the giant hummingbird, *Patagona gigas*. At 8½ inches (21 cm) it is huge compared with the bee hummingbird, *Mellisuga helenae*, of Cuba, which is little more than 2 inches (5 cm) long. Half of the bee hummingbird's length is bill and tail, the body being the same size as a bumblebee. It is the world's smallest living bird.

Hummingbirds are very diverse in form, although all of them are small and have the characteristic rapid wingbeats producing the hum that gives them their name. They have brilliant, often iridescent, plumage, which has led to their being given exotic names such as ruby and topaz, and also to their being killed in their thousands and their skins exported to Europe for use in ornaments. A feature of many hummingbirds is the long, narrow bill, often straight but sometimes curved, as in the two species of sicklebills in the genus *Eutoxeres*. The sword-billed hummingbird, *Ensifera ensifera*, has a straight bill as long as its head, body and tail put together. Hummingbirds are most common in the forests

of South America, but they range from southern Alaska to Tierra del Fuego, the southernmost tip of South America. Some species are so rare that they are known only from museum collections.

Hummingbird stamina

Considering the diversity of habitats and food in the South American forests, it is not surprising that there should be so many kinds of hummingbirds living there. It is rather surprising, however, to learn that hummingbirds breed as far north as southeastern Alaska or in such high altitudes as the heights of the Andes. The rufous hummingbird, *Selasphorus rufus*, breeds from Alaska south to California, migrating to Mexico for the winter, a very long journey for so small a bird.

The ruby-throated hummingbird, *Archilochus colubris*, also migrates to and from North America, where it is found from Nova Scotia south to Florida. It crosses the Gulf of Mexico or passes through Mexico overland on each trip. Unlike nonmigratory hummingbirds, it stores a layer of fat equal to half its body weight before setting off. This store amounts to less than 1 ounce (28 g), and the distance a bird can fly is propor-

Rufous hummingbirds (male, below) are the northernmost species, breeding as far north as southeastern Alaska.

RUBY-THROATED HUMMINGBIRD

CLASS	**Aves**
ORDER	**Apodiformes**
FAMILY	**Trochilidae**
GENUS AND SPECIES	***Archilochus colubris***

WEIGHT
Up to 2 oz. (56 g)

LENGTH
Head to tail: 3½ in. (9 cm)

DISTINCTIVE FEATURES
Tiny size; compact, strongly muscled body; long, extremely thin bill; forked tail.
Male: metallic green upperparts; iridescent red throat, appears black in poor light; whitish underparts. Female: dusky white chin and throat; buff flanks.

DIET
Flower nectar, small arthropods and flies

BREEDING
Age at first breeding: 1 year; breeding season: April–June; number of eggs: 2; incubation period: 16 days; fledging period: 15–20 days; breeding interval: 2 or 3 broods per year

LIFE SPAN
Up to 9 years

HABITAT
Breeding: deciduous and mixed forest; also parkland and gardens. Winter: tropical forest and scrub.

DISTRIBUTION
Breeds in eastern North America from Florida and Louisiana north to Alberta and Nova Scotia. Winters in Central America from central Mexico south to Costa Rica.

STATUS
Common in southern half of breeding range, becoming uncommon farther north

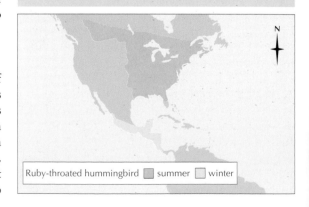

Ruby-throated hummingbird ▢ summer ▢ winter

A hummingbird's wings are connected only at the shoulder joint and thus can move in all directions, allowing the bird to hover while feeding. A male ruby-throated hummingbird is shown.

tional to the ratio of fuel and body weight. Small birds can carry a large store of fat for their size and so are capable of great feats of stamina. Research suggests that ruby-throated hummingbirds carrying at least ¼ ounce (2 g) of fat should be able to fly more than 590 miles (950 km), far enough for a nonstop flight across the Gulf of Mexico. Hummingbirds also have, relatively, the largest pectoral (breast) muscles of any birds, up to 30 percent of their total weight.

Hummingbird speed

Even ignoring these migratory feats, the flight of hummingbirds is truly remarkable. Their wings beat so fast they appear as a blur. Small species have wingbeats of at least 80 per second and in courtship displays even higher rates have been recorded. There are reports, as yet unconfirmed, of wingbeats of up to 200 per second. The fast wingbeats enable the hummingbirds to dart to and fro, jerking to a halt to hover steadily. They are also extremely fast in straight flight and speeds of 70 miles per hour (112 km/h) have been recorded. In dives they can reach 60 miles per hour (95 km/h). Specialized filming has shown that hummingbirds do not take off by leaping into the air like other birds, but instead lift off using their rapid wingbeats.

Nightly hibernation

Flying in this manner requires a large amount of energy, so hummingbirds must either feed constantly or have plentiful energy reserves. Even at rest their metabolism (the rate at which they produce energy) is 25 times faster than that of a chicken and their heart rate might reach 1,260 beats per minute. At night, when they are unable to feed, hummingbirds conserve their food reserves by becoming torpid (going into a form of nightly hibernation). In the Andes Mountains a hummingbird's temperature drops from 100° F (38° C) to 66° F (19° C), about the temperature of the surrounding air, and their metabolism is reduced six times.

Nectar seekers

Hummingbirds feed on nectar and small soft-bodied animals. To sip nectar they hover in front of flowers and insert their pointed bills down the corolla (the petals' base) or, if that is too long, pierce it near the bottom. The nectar is sucked through a tubular tongue that resembles those of flowerpeckers. Pollen is often brushed onto the hummingbirds' heads and transferred to other flowers, so pollinating them. For the flowers of the South American jungle, hummingbirds have an extremely important role as pollinators.

Small insects are caught on the wing and spiders are taken from their webs. Most species of hummingbirds are unable to manipulate insects in their bills and have to rush at them, forcing the prey into their bills. Some pick insects and spiders from flowers. Individual hummingbirds often consume more than half their total weight in food, and may drink twice their weight in water, every day.

Aerobatic displays

In most species the males defend a territory and during courtship they display to the females, flying about in arcs, swoops and dashes, singing songs that are almost too high-pitched for humans to hear. Most hummingbirds have twittering or squeaky songs. The males are usually

promiscuous, mating in the air with several females, but in a few species, such as the violet-eared hummingbirds (*Colibri* spp.) pair bonds are formed and the male helps rear the family.

In most hummingbirds the nest is a delicate cup of moss, lichen and spiderwebs placed on a twig or among foliage. The two eggs are incubated for 2 or 3 weeks and minute, naked chicks hatch out. They are fed by the parent hovering alongside, putting its bill into theirs and pumping out nectar. The chicks grow very rapidly and leave the nest by about 3 weeks old.

Hovering skill

When feeding, hummingbirds can be seen hovering steadily and even flying backward. They can do this because their wings are connected to the body only from the shoulder joint, and therefore can swivel in all directions. When hovering, the body hangs at an angle of about 45° so the wings are beating backward and forward instead of up and down.

In each complete beat the wing describes a figure eight. As it moves forward (the downstroke), the wings are tilted so they force air downward and the bird upward. At the end of the stroke the wings flip over so that the back is facing downward and on the upstroke air is again forced downward. To fly backward, the wings are tilted slightly so air is forced forward as well, and the hummingbird is driven back.

A hummingbird is able to consume more than half its total body weight in nectar and insects each day.

HUMPBACK WHALE

Mother with her calf. Individual humpbacks can be told apart by the distinctive white markings or patches on their bellies and flippers and beneath their tail flukes.

THE HUMPBACK WHALE belongs to the same family as the better-known baleen whales, but has a number of distinct features that place it in a genus of its own. Its name probably comes from its appearance as it dives, when it arches its back just before disappearing below the water's surface. This whale's most characteristic feature is an extremely long set of flippers. Far longer in proportion than in any other whale, they may be as much as one-third of its total body length. The body, far from being streamlined, is quite barrel-shaped and stocky.

Individual markings

Humpbacks migrate to colder climates to feed in summer. They tend to hug the coastlines as they pass the Tropics to waters of high latitudes. Keeping close to the coasts in this way has made their migrations well-known but has also made the whales an easy prey for hunters in the past.

Humpbacks grow 38 to 50 feet (11.5–15 m) in length, the females being slightly longer than the males. Their color is normally black above and white below, but there are a number of variations on this. The back is generally all black or a very dark, slate gray, but the underside varies considerably from almost totally white to nearly all black. Some scientists have tried to divide humpbacks into subspecies on the basis of their color, but this method has been found to be unreliable.

However, it is probable that there is some sort of division, as before humpbacks were protected, whalers would notice they caught all-dark whales, then all-light whales as they worked through various schools passing along the coast. There are also sufficient color variations for these to be used in recognizing individual animals.

Long, distinctive flippers

The humpback's long flippers are usually dark on the upper or outer surface and white on the lower or inner surfaces. In the same way the tail flukes are dark above and pale below. The flippers also have a distinctive outline. The lower margins are scalloped, and they have a number of irregular humps or tubercles along this edge. These tubercles also occur on the upper part of the head and along the jawline, and each usually has one or two short coarse hairs growing from it.

On its underside, the humpback has a number of grooves running as far back as the navel. These number 2 to 36 compared with an average of 85 to 90 in the fin and blue whales. Each groove is separated from its neighbor by as much as 8 inches (20 cm), and sometimes the concave part of the groove contrasts with the body color around it. The whale's short dorsal fin is set rather far back and is stubby with a broad base. There may be as many as 400 baleen (whalebone) plates on each side of the whale's

HUMPBACK WHALE

CLASS	**Mammalia**
ORDER	**Cetacea**
FAMILY	**Balaenopteridae**
GENUS AND SPECIES	***Megaptera novaeangliae***

ALTERNATIVE NAME
Humpbacked whale

WEIGHT
Up to 53 tons (48 tonnes)

LENGTH
38–50 ft. (11.5–15 m)

DISTINCTIVE FEATURES
Stocky, barrel-shaped body; extremely long flippers; low, stubby dorsal fin; tubercles (humps or knobbles) on head, lower jaw and lower edge of flippers; black or dark gray overall, with white patches of varying size on belly and beneath tail flukes

DIET
Southern humpback: mainly krill.
Northern humpback: krill; small fish such as mackerel, anchovies, sardines and capelin.

BREEDING
Age at first breeding: 4 or 5 years; breeding season: primarily late winter; gestation period: 330–345 days; number of young: 1; breeding interval: 2 years

LIFE SPAN
Up to 70–80 years

HABITAT
Seas and oceans; mainly in coastal waters, including small bays and estuaries

DISTRIBUTION
Summer: high-latitude polar waters.
Winter: warmer waters near equator.

STATUS
Uncommon; estimated population: 12,000 to 15,000

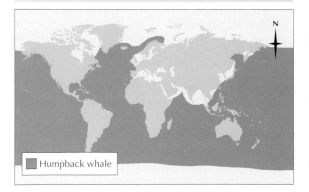
Humpback whale

It has been suggested that humpbacks leap from the water so as to rid themselves of barnacles. This whale was photographed off Admiralty Island, Chatham Strait, Alaska.

mouth, but the average is around 300. Each plate is 2 feet (60 cm) in length and grayish black in color. Sometimes there are a few white baleen plates. When present, these are usually at the front of the mouth and they are often associated with blotchy white markings on the skin in about the same position as the plates themselves.

Summer and winter migrations

Humpbacks are found in all oceans but are typically whales of the coasts, often coming close inshore, even into small bays and estuaries. In spite of this, they are very rarely found stranded as some other whale species often are.

Humpback whales migrate every summer to polar waters at high latitudes to feed. They then migrate back to warmer, tropical waters in winter. It is at this time that the young are born and mating takes place. Although there are separate populations of humpbacks in both Arctic and Antarctic waters, by moving toward the equator in the winter months there is possibly some interchange between the two.

The humpbacks sometimes seen off the British coasts spend the summer months feeding with their calves to the north and east of Norway. They then move westward and then south in February and March, migrating as far as the western coast of North Africa. Here another generation of calves is born, and further mating takes place during April and May. After this the whales migrate north again, passing the Outer Hebrides and the Faeroes, to the north of Scotland, and finally reaching northern Norway around July or August.

The migrations of the southern populations of humpbacks have been studied in considerable detail, and these follow the same pattern as those in northern waters. The whales spend the summer in the Antarctic feeding on the abundant krill and some small fish. As the winter approaches they move gradually northward.

Orderly migrations

The first southern humpbacks to go north are the females that have just finished suckling their calves. The newly weaned calves go with them. Next are the immature animals, then the mature males and finally the pregnant females. They all go as far as the warm equatorial waters, where the pregnant females give birth and then mate once more. In the return migration the pregnant females go first, followed by immature animals, then the mature males and finally the adult females with their newborn calves. By the time they reach the Antarctic feeding grounds the herds have all mixed together and they stay this way until it is time to travel north again.

Krill is the main food

The food of humpback whales, particularly those that are found in the Southern Ocean, consists mainly of krill. Whales in the Northern Hemisphere are more likely to supplement this diet of planktonic crustaceans with a variety of small fish such as mackerel, sardines, anchovy and capelin. When the humpbacks are in tropical waters to breed, they actually feed very little. Most of the feeding is done in colder waters, where enough reserves of blubber are built up to last through the rest of the year.

Mating antics

Humpbacks are well known for their amorous antics. They roll over and over in the water, slapping the surface or each other with their long flippers. This causes considerable commotion and the noise is said to be audible several miles away. Sometimes the whales leap completely clear of the water in their play, although it has been suggested that this is more often done to rid themselves of encrusting barnacles.

Killed on the coast

As is the case with many whale species, the humpback's greatest enemies used to be commercial whalers. Killer whales or orcas take their usual toll, but the humpback, with its coast-hugging habits and fixed migratory routes, was an easy prey for humans. It used to be that when a whale fishery started up, it was usually the humpback species that was killed first. Today there is total protection of humpbacks apart from a few subsistence whaling operations in Greenland, Tonga and the Caribbean. Their numbers have now risen to about 12,000 to 15,000 whales.

Barnacle trouble

Humpback whales are usually heavily infested with barnacles and whale lice. It was long ago noted that humpbacks passing the South African coast on their way north were heavily barnacled but those returning from tropical waters were only lightly infested. It was believed that when the whales got to where the Congo River emptied into the sea they moved inshore into much less salty water, where the barnacles died and dropped off.

A humpback whale dives off the coast of southeastern Alaska. Humpbacks migrate to high latitudes in summer to feed.

HUNTING WASP

WASPS CAN BE DIVIDED by their habits into social and solitary types. Social types are the familiar wasps and hornets, which live in colonies or communal nests, may sting severely and often appear very numerous. They are, however, only numerous in terms of individual insects. A far greater number of species of wasps are solitary types, in which each individual leads an independent life.

Apart from a few exceptions, solitary wasps are hunters. The females hunt other insects and store them in burrows or hollow mud cells. This is not to provide food for themselves, but as a store for their larvae. The exceptions are the solitary wasps that have a parasitic or "cuckoo" mode of breeding. They lay their eggs in the burrows of other solitary wasps, or solitary bees. Their larvae then grow by feeding on the store provided or by killing and devouring the larvae for which this store was intended. A hunting wasp, then, is a solitary wasp that has not adopted a parasitic mode of life.

Hidden larders

Female hunting wasps spend most of their time providing food and shelter for their offspring. There is a great variation from one species to another in the sequence and details of reproduction, but the general pattern is similar.

After mating the female digs a burrow in earth or rotten wood, or seeks out a hollow plant stem, or constructs a receptacle of mud, plastered on while wet and allowed to dry. She then hunts for living insects. The type and number of insects depends on the species of wasp. When one is found, for example a caterpillar, it is stung, but the effect of the injected poison is only to paralyze it, not to kill it. The caterpillar is then carried to the burrow and put inside, and other caterpillars are sought out, stung and added to the first. When sufficient numbers have been caught, the wasp lays an egg, usually on the wall above the immobilized victims, seals up the burrow and goes on to build and stock similar burrows.

Species such as *Eumenes coarctatus* and *Ammophila pubescens* both hunt caterpillars and stock their burrows in this way. Other species place only one insect in each burrow. For example the spider-hunting wasps of the family Pompilidae usually catch only one spider per burrow.

After a short time the egg hatches and the hunting wasp larva feeds on the store of living food, sucking the juices of the insect or spider in such a way that it remains alive until almost completely consumed. The residue of skin quickly

shrivels and dries and the wasp larva goes on to its next victim. Most overwinter as mature larvae. They then pupate and adults emerge from the burrows the following spring. Some species of the family Sphecidae show a degree of parental care by remaining with the developing larvae and providing food at appropriate times.

Perfect meat stores

The provision that the female wasp makes for her offspring is quite elaborate. She constructs a shelter to protect it from enemies and from extreme temperatures, and she provides a store of food that is kept not merely fresh, but alive. There is therefore no danger of the uneaten part of the store putrefying during the development of the wasp larva. Moreover, the tiny, delicate larva cannot be injured by the protesting struggles of its victims, for they are paralyzed.

Three families

Hunting wasps are placed in three families: Pompilidae, Eumenidae and Sphecidae. To a certain extent they can be grouped depending on their prey type. Pompilids hunt spiders, whereas the other two families feed on a range of insects.

Members of the Sphecidae, or digger wasps, by far the largest group of solitary wasps, construct a variety of burrows and some nest in wood. A number of different insects are attacked,

Female spider-hunting wasps often take spiders several times their size, paralyzing their victims before taking them to be used as a food store for their offspring.

HUNTING WASPS

PHYLUM **Arthropoda**

CLASS **Insecta**

ORDER **Hymenoptera**

SUPERFAMILY **Pompiloidea**

FAMILY **Spider-hunting wasps, Pompilidae; digger and sand wasps, Sphecidae; potter and mason wasps, Eumenidae**

SPECIES **16,000 to 19,000**

ALTERNATIVE NAME
Digger wasps (Pompilidae only)

LENGTH
Most species: ⅖–⅘ in. (1–2 cm)

DISTINCTIVE FEATURES
Chewing mouthparts; 2 pairs of transparent wings; highly developed ovipositor (egg-laying tube) in female. Pompilidae: many species orange and black with long legs. Sphecidae: club-shaped abdomen.

DIET
Pompilidae: spiders (female and larva); flower nectar (male). Sphecidae: insects such as aphids, flies, beetles and froghopper nymphs. Eumenidae: mainly caterpillars.

BREEDING
Varies according to species. Number of eggs: 1 per burrow; larval period: most species overwinter as larvae.

LIFE SPAN
Several weeks to 1 year

HABITAT
Varies; often in areas with sandy soils

DISTRIBUTION
Virtually worldwide, mainly in warm regions

STATUS
Generally common

In common with other members of the family Eumenidae, potter wasps stock their nests with caterpillars to feed their young.

but usually each species is relatively specific in its prey choice. Thus members of the genus *Ammophila* dig burrows in sandy soil and stock them with caterpillars. The "bee killers" of the genus *Philanthus* make a similar type of nest but stock it with bees. *Gorytes*, also burrowers, drag young froghoppers out of their concealing mass of secreted foam. The mainly tropical mud-dauber wasps, of *Sceliphron* and allied genera, make nests of mud plastered onto any suitable surface, including the walls of houses.

One British wasp of the family Eumenidae is the potter wasp (*Eumenes* sp.), named for the clay nests it builds. These look like tiny, round flasks, each with a short neck and flared rim. The nests are stocked with small caterpillars and can be found on heather on dry sandy heaths. Potter wasps are less common than they were because their heathland habitats are being destroyed.

Victory against heavy odds

The most spectacular of the hunting wasps are the spider hunters of the family Pompilidae. Some of the tropical pompilids are among the biggest of all wasps and they hunt the large bird-eating spiders and tarantulas. These formidable spiders, which can kill a mouse with ease, live in burrows that the female wasp will sometimes enter. When they fight in the open, the two may spar around or they may close and tumble like wrestlers. The wasp's only weapon is her sting, which is pitted against eight clutching legs and a pair of powerful poison fangs. Almost invariably, however, the wasp is the victor, and manages to slip her sting into the undersurface of the spider, where its nerve centers lie. Some spider-hunting wasps appear to deliver the first sting into the spider's head to immobilize its jaws and then close in to finish the battle.

When large prey of this kind is chosen, the wasp has considerable difficulty in carrying it. In such cases the usual procedure is for the wasp to find her victim, sting it and then look for a convenient place nearby to make a burrow. When this is ready, the victim is dragged to it and pushed in, and an egg is then laid.

There are other hunting wasps, for example some of the wasps in the superfamily Scolioidea. These do not transport their prey to a burrow, but paralyze it and lay one or more eggs on the prey item, wherever it happens to be found.

HUTIA

UTIAS ARE LARGE, ratlike rodents that resemble the coypu or nutria, *Myocastor coypu*, and the hyraxes (family Procaviidae). Most people outside the islands of Central America and the Caribbean have never heard of hutias, but they have a remarkable history. The coypu is a semiaquatic rodent of South America and the chances are that the ancestors of hutias were also South American. Indeed, there was a Venezuelan hutia, *Procapromys geayi*, but only one specimen is known and even this is believed to have been a young individual of the hutia living on the Isle of Pines, off Cuba. Apart from this, there used to be around 25 species of hutias on Cuba, Haiti, Jamaica, the Bahamas and other islands lying between North America and South America. Of these 25 species, 18 are now extinct, having been wiped out during the last few centuries. Some had just become extinct when Europeans first reached the New World. The few species still surviving are already becoming rare and most are in danger of being wiped out.

Hutias look very like the hyraxes of Africa, although the two groups are not closely related and hyraxes are not even rodents. Nonetheless, both have the local name of coney. They have the blunt muzzle of the coypu and its coarse coat, but in most hutias the tail is not so long as in the coypu. Hutias have small eyes and ears and are yellowish to reddish brown, black or gray in color. The four species of hutias on Cuba and the Isle of Pines are up to 1¾ feet (50 cm) long with a tail of up to about 1 foot (30 cm) in length, and they can weigh nearly 19 pounds (8.5 kg).

Some in trees, others among rocks

Hutias are omnivorous, feeding on fruits, leaves, bark, lizards and any other small animals they can find. They also eat grass and browse shrubs. Cuban hutias, *Mesocapromys* and *Mysateles*, are diurnal (day-active) and live in trees. When in danger they come to the ground, seeking safety in holes.

Jamaican and Bahamian hutias, *Geocapromys*, are terrestrial and mainly nocturnal. They are slightly smaller than those hutias already mentioned and live on broken ground among rugged limestone hills. They feed on grass, leaves and fruit. Two related species are known, one on Cuba from bones found in caves and one on the Bahamas that became extinct many centuries ago, probably killed off by prehistoric Stone Age hunters.

Little is known of the Hispaniolan hutias, of the genus *Plagiodontia*, although they are thought to be mainly active at night.

Female hutias give birth to between one and three young after a gestation period of around 120 days. In the case of the Cuban hutias, the young are born with reddish brown fur, which later turns gray. The young hutias can run about soon after birth. Breeding is year-round in the wild and the female might have two or more litters each year.

More extinct than living

Hutias have probably been hunted by humans as a food source for hundreds of years. One of the first species to become known to scientists was discovered from bones found in caves. Today most of the species that are known to scientists are extinct and the bones of these are often found in kitchen middens, the waste pits of long ago. The remains of long-extinct species of hutias have been found on Haiti, the Dominican Republic, Puerto Rico, the Virgin Islands and elsewhere.

Of the 25 known species of hutias, 18 are now thought to be extinct. Most of those surviving are threatened by hunting, habitat loss and introduced animals such as mongooses.

Cuban hutias are arboreal (tree-living), mainly feeding on fruits and leaves, but also hunting small prey such as lizards.

Cuban hutias, meanwhile, are still hunted today with dogs for their flesh. Of these the bushy-tailed hutia, *Mesocapromys melanurus*, and the dwarf hutia, *M. nanus*, are listed as being rare. The Jamaican and Bahaman species, meanwhile, are now very rare but the few survivors continue to be hunted for food and sport. Those that are critically endangered include Cabrera's hutia, *M. angelcabrerai*, and the large-eared hutia, *M. auritus*. Other species are vulnerable or lower risk, but destruction of habitat is also a problem, as is predation by mongooses, introduced into some areas from India.

Each island its own species?

Already a picture can be built up of the islands being inhabited by many species of hutias, different species in different areas of the Caribbean, probably all descended from coypu-like ancestors from South America. They were probably very numerous thousands of years ago, with the giant barn owl their only predator until humans arrived. As the hutias grew rare, hunted by primitive humans, the giant barn owl probably died out and now the hutias also appear to be dying out, killed first by humans for food, and now by introduced mongooses and rats, and by habitat destruction. Every few years the bones of yet another extinct species of hutia are found, showing that formerly these animals existed in many more species than the 25 biologists are already aware of.

The Caribbean has been the scene of several notable introductions of animals, most of them without thought about the effect on indigenous wildlife. Many of these have disturbed, if not destroyed, some of the local animals. Examples are the brown or common rat, *Rattus norvegicus*, mongooses and domestic cats and dogs.

HUTIAS

CLASS	**Mammalia**
ORDER	**Rodentia**
FAMILY	**Capromyidae**
GENUS	**Small Cuban hutias, *Mesocapromys*; long-tailed Cuban hutias, *Mysateles*; Jamaican and Bahamian hutias, *Geocapromys*; Hispaniolan hutias, *Plagiodontia*; others**
SPECIES	**Probably 7, including dwarf hutia, *Mesocapromys nanus*; Garrido's hutia, *Mysateles garridoi*; and Cuvier's hutia, *Plagiodontia aedium***

WEIGHT
Up to 19 lb. (8.5 kg)

LENGTH
Head and body: 8–20 in. (20–50 cm); tail: 1⅓–30 in. (3.5–30 cm)

DISTINCTIVE FEATURES
Large, bulky rodent resembling coypu or hyrax; blunt muzzle; small eyes set back on head; small ears; coarse coat

DIET
Grasses, shoots, leaves, fruits, bark and small animals, including lizards

BREEDING
Breeding season: all year; gestation period: about 120 days; number of young: 1 to 3; breeding interval: 2 or more litters per year

LIFE SPAN
Up to 10 years in captivity

HABITAT
Forests, plantations and rocky areas

DISTRIBUTION
Central America; Caribbean islands including Jamaica, Bahamas, Dominican Republic, Hispaniola, Cuba and Haiti

STATUS
Critically endangered: 2 species; vulnerable or lower risk: 5 species

Hutias

HYDRA

THE HYDRA'S SIMPLE, tubular body with its crown of tentacles has made it the object of many detailed studies. It is one of the few freshwater cnidarians, the bulk of which are marine. The body of the hydra is a bag, the wall of which is made up of two layers of cells separated only by a very thin layer of noncellular material. Its tentacles, usually between 4 and 12 in number, are hollow. They surround the mouth, while the other end of the body is a basal disc, which normally anchors the hydra by a sticky secretion. Although often abundant in ponds, hydras frequently escape notice because of their habit of retracting into a tiny blob when disturbed.

Both the hydra's tentacles and body are very extensible, for the bases of many of the cells are drawn out like muscle fibers. Those of the outer layer of cells run lengthwise and those of the inner layer run around the body. This allows the hydra to contract its body periodically. It coordinates its movements through an extremely simple nervous system, consisting only of a network of nerve cells. There is no brain of any sort.

Colors caused by algae

Two common species found worldwide are the green hydra, *Chlorohydra viridissima*, and the brown hydra, *Hydra (Pelmatohydra) oligactis*. The green hydra has short tentacles that are never as long as its body. The brown hydra, on the other hand, has tentacles four or five times the length of its body which is usually clearly divided into a stomach region and a narrower stalk. Their colors are caused by single-celled algae living within their cells. When animal prey is scarce, the hydra draws nourishment from these algae. In both species the body may be as much as 1⅕ inches (3 cm) long, but it is usually much shorter, about ⅖ inch (1 cm).

The stinging cells

Hydras, like their relatives the sea anemones and jellyfish, have stinging cells with which they capture their prey. Each stinging cell or nematocyst is a rounded cell with a hollow coiled thread inside that can be shot out at great speed. The hydra has four kinds of nematocysts. In one the thread is shot into the prey, injecting a poison. In a second kind the thread coils after it is shot out, and if the prey has bristles of any kind, they often become entangled. The third type of nematocyst is probably truly defensive. It is shot out at animals not normally eaten by the hydra. The fourth kind is used to fasten the tentacles when the hydra is walking. This is not strictly a stinging cell, and is best referred to as a thread capsule. In fact, some people prefer to use the term "thread capsule" for all of these cells simply because some of them do not sting. In all types, the nematocysts cannot be used more than once but are replaced by new ones migrating in from other parts of the body.

Progressing by somersaults

Although normally anchored, the hydra can move about by creeping slowly on its basal disc in a sort of sliding movement. It can move more rapidly, however, by a looping movement that looks like a series of somersaults. To do this, the hydra bends over and gets a grip with special thread capsules on the tentacles. It then lets go with its basal disc and brings the disc over as if it were doing a cartwheel. Hydras can also float at the surface of the water, buoyed up by gas bubbles given out by the basal disc.

Snagging its prey

Hydras feed on insect larvae, water fleas, worms and even newly hatched fish and tadpoles. Between meals the tentacles are held outstretched and more or less still, but at the approach of prey they start to writhe and later they bend in toward the open mouth. A single chemical, glutathione, given out by the prey causes this reaction. Then, if the prey touches the tentacles,

Hydra vulgaris budding, that is, reproducing asexually by forming buds that break away. It has been argued that because of this method of reproduction, hydras are in fact immortal.

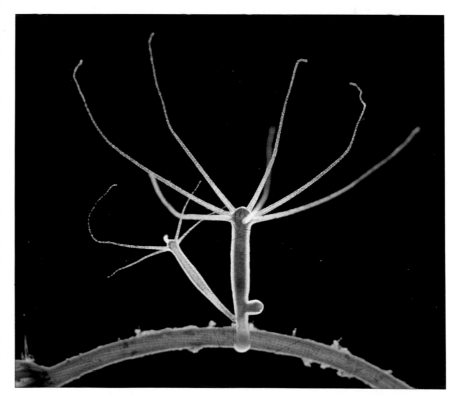

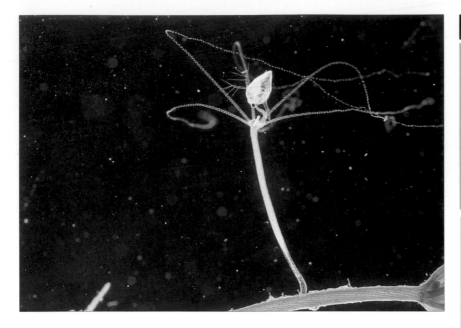

Here a **Hydra vulgaris** captures a daphnia, a minute freshwater crustacean. Animals are detected because they secrete a chemical into the water. They are then poisoned and caught using threads from the nematocysts, or stinging cells.

the threads of the nematocysts are shot out, catching the animal. It is held and paralyzed, then carried to the mouth and swallowed. The mouth can open wide enough to take in animals that are themselves wider than the body of the hydra, which will stretch to accommodate them.

Once inside the baglike body of the hydra, the prey is partially digested by enzymes given out by the inner layer of cells. Small particles breaking off are engulfed by individual cells for the final stages of digestion, and indigestible particles are rejected through the mouth.

The immortal hydra

Hydras reproduce both sexually and by budding. Budding is more common, with each bud beginning as a little bump on the side of the hydra's body. This grows out, and an opening appears at its free end. Tentacles are pushed out all around the mouth, and finally the bud breaks away from the parent, settles down and grows into a new hydra, the whole process taking just 2 days. A single hydra may bear several buds at once. Because this is the main method of reproduction, it has been argued that hydras are immortal.

Hermaphrodites

Most species of hydras reproduce sexually only in response to harsh conditions. Such conditions include cold or hot climates, drought and chemical pollution including an accumulation of carbon dioxide in the water. This last happens if a population of hydras becomes overcrowded. Offspring produced by sexual reproduction are more mobile, and can migrate to colonize new habitats where conditions are better.

There are no special reproductive organs, but small cells appear as bulges on the upper half of the body. Most species are at least capable of

HYDRAS

PHYLUM	**Cnidaria**
CLASS	**Hydrozoa**
ORDER	**Hydroida**
FAMILY	**Hydridae**
GENUS AND SPECIES	**Green hydra, *Chlorohydra viridissima*; brown hydra, *Hydra oligactis*; *H. vulgaris*; others**

LENGTH
Most species: up to ⅖ in. (1 cm); largest species: up to 1⅕ in. (3 cm)

DISTINCTIVE FEATURES
Simple, hollow bag comprising 2 layers of cells; 4 to 12 tentacles around mouth; able to contract; can regenerate from a small fragment

DIET
Small freshwater animals such as insect larvae, crustaceans, worms, tadpoles and fish larvae

BREEDING
Usually reproduces asexually by forming buds that break away (takes 2 days); sexual reproduction takes place only in harsh conditions; embryo may lie dormant for 3–10 weeks

LIFE SPAN
Indefinite, as "budding" is ongoing; all cells replaced about every 45 days

HABITAT
Freshwater rivers, ponds and lakes

DISTRIBUTION
Worldwide

STATUS
Abundant

being hermaphrodites. Ovaries might appear lower down on the body, also as bulges, each containing a single, large egg cell, or ovum. The ripe ovum pushes through the outer layer of the hydra's body, and the cells around it form a little cup or cushion for the ovum. The male sperms are shed into the water, where they swim about and eventually reach the ova and fertilize them. The embryo that results from the division of the fertilized ovum secretes a hard, sticky yellow shell around itself. The shell may be smooth or spiny, according to the species. Thus enclosed, the embryo can survive harsh conditions such as drying and freezing. After lying dormant for 3–10 weeks it breaks out of its capsule, grows tentacles and becomes a new hydra.

HYENA

THERE ARE THREE SPECIES of hyena: the spotted or laughing hyena, *Crocuta crocuta*, of sub-Saharan Africa; the brown hyena, *Parahyaena brunnea*, of southern Africa; and the striped hyena, *Hyaena hyaena*, which ranges from northern and northeastern Africa through Asia Minor to India. They all have massive heads and powerful jaws and teeth with which they can crack even marrow bones. Their ears are large and their shoulders are markedly higher than their hindquarters. Their tails are quite short, about 1 foot (30 cm) long, and each foot has five heavily padded toes. The male spotted hyena may be 6 feet (1.8 m) long in its head and body, 3 feet (90 cm) high at the shoulder and can weigh up to 200 pounds (90 kg). The female is slightly smaller than the male.

Distinctive coats

The three species of hyena are easily recognized by their coats. That of the brown hyena is dark brown with a grayish neck and lower legs. The coat is exceptionally long and heavy. The striped hyena has a gray to yellowish brown coat with brown or black stripes. It has strong forepaws, which are well adapted for digging up meat from caches made by other carnivores. Both the brown and striped hyenas have long-haired manes that can be erected. The spotted hyena is the largest species of hyena. It has a gray to tawny or yellowish buff coat with many brown or black spots. It has only a slight mane and shorter ears than the other species. Its jaws are probably the most powerful in proportion to overall body size of any living mammal.

Spotted hyena clans

Hyenas are found on arid, open scrubland and savanna. They spend most of the day in holes in the ground, in caves or in lairs in dense vegetation. Although typically nocturnal, hyenas are sometimes active by day. Spotted hyenas live in clans of up to 100 animals at times, and have well-defined territories marked by their urine

The spotted or laughing hyena is the largest and most aggressive hyena. It has powerful jaws and strong teeth and will eat the entire remains of a kill.

Striped hyenas are found in India and Asia Minor as well as Africa. This female and her cub were photographed in Israel.

HYENAS

CLASS	**Mammalia**
ORDER	**Carnivora**
FAMILY	**Hyaenidae**

GENUS AND SPECIES **Spotted hyena, *Crocuta crocuta*; brown hyena, *Parahyaena brunnae*; striped hyena, *Hyaena hyaena***

ALTERNATIVE NAME
Spotted hyena: laughing hyena

WEIGHT
65–200 lb. (30–90 kg)

LENGTH
Head and body: 3–6 ft. (0.9 m–1.8 m); shoulder height: 2–3 ft. (60–90 cm)

DISTINCTIVE FEATURES
Massive head with blunt muzzle and strong jaws; muscular neck; shoulders higher than hindquarters; shaggy coat with variable pattern of spots or stripes; short, hairy tail

DIET
Scavenge other carnivores' kills; also take rodents, insects, eggs, fruits, plant matter and (spotted hyena only) larger mammals

BREEDING
Age at first breeding: 2–3 years; breeding season: May–August; gestation period: 88–110 days; number of young: 1 or 2 (spotted hyena), 2 to 4 (brown and striped hyenas); breeding interval: 12–40 months

LIFE SPAN
Probably up to about 20 years

HABITAT
Arid, open scrubland and savanna

DISTRIBUTION
Spotted hyena: sub-Saharan Africa. Brown hyena: southern Africa. Striped hyena: North Africa, through Middle East to India.

STATUS
Populations of all species declining

and droppings. Members of other clans are driven from the territory. They often hunt in packs, are able to run at up to 40 miles per hour (64 km/h) and are more aggressive than the other two species. The voice of the spotted hyena is a howl, made with the head held near the ground, beginning low but becoming louder as the pitch rises. When excited, as during the breeding season, this hyena makes its so-called laugh. This is a shrill social-appeasement call that sounds like a maniacal human giggle.

Kill as well as scavenge

Contrary to the popular belief that all hyenas are scavengers, the diet of spotted hyenas mainly consists of their own mammalian kills, especially zebra and antelopes such as wildebeest. They may also take domestic animals. Nonetheless, this species will sometimes scavenge from the kills of other large carnivores such as lions. Rather than waiting for the remains of such a kill, the hyenas might form a pack and rush on the lion, driving it away. Only around 30 percent of the spotted hyena's diet is made up of scavenging, however. They will also eat some small insects and may even feed on locusts. Immensely strong in the jaws and shoulders, the spotted hyena is said to be able to carry away a human body or the carcass of an ass.

Sexes look alike

Spotted hyenas have very similar-looking reproductive organs and it is difficult to distinguish the sexes, with both appearing to be male. Mating takes place between May and August and the gestation period is around 110 days. At birth the mother takes up a squatting position, and the young are ejected forward. She usually gives birth to one or two babies, which are born

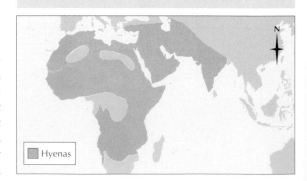

Hyenas

fully furred. Although hyenas have been known to live up to 40 years in captivity, their life span is considerably less in the wild, perhaps 20 years.

Striped hyenas

The striped hyena is usually smaller than the spotted hyena. Its diet differs, too, in that it rarely kills its own mammalian fodder and relies more on scavenging the kills of other carnivores. It also eats small vertebrates, insects and fruits.

Breeding also differs from that of the spotted hyena in that the gestation period is 90 days and there are two to four young in the litter, sometimes five. Otherwise it is much the same. The babies are born in holes in the ground, blind and with their ears closed.

Brown hyenas

The brown hyena is midway in size between the spotted and striped hyenas. Its breeding and feeding habits are similar to those of the striped hyena. Some brown hyenas live near the seashore and feed on carrion left by the receding tide, eating anything from dead crabs to the carcasses of stranded whales. For this reason the species was once known as the strandwolf.

Reputation for cunning

The striped hyena has given rise to all manner of beliefs about its magical powers and its cunning. One possible cause of such ideas is its habit of "playing dead" when cornered by dogs or other predators. The hyena lies perfectly still, and the predators, having sniffed around it for a while, lose interest and turn away. At that moment the hyena jumps to its feet and dashes away.

Not so cowardly

Another belief that has persisted to this day is that hyenas are cowardly because they live by feeding on the remains of lions' and other large carnivores' prey items. This reputation was proved unfounded, at least in the case of the spotted hyena, some years ago.

One of the first studies to cast doubt on the spotted hyena's cowardly reputation was carried out by Dr. Hans Kruuk, of the Serengeti Research Institute in Arusha, Tanzania. He found that at dusk they came out of their holes, meeting other hyenas and exchanging greeting ceremonies until a pack of up to 20 animals was formed. Then they set off at a seemingly leisurely pace until they found a family party of zebras or a herd of wildebeest. They began to harass their quarry, snapping at them until they had slowed one down, at which point all the hyenas concentrated on the single victim. By dawn not even a splinter of bone would be left. Kruuk found that although spotted hyenas are scavengers, they go out to kill for themselves. In fact, as often as not it was the lions that partook of the spotted hyenas' kills instead of the other way around.

Despite their reputation for scavenging the kills of other carnivores, spotted hyenas kill most of their own prey, such as this adult zebra.

HYRAX

A Cape hyrax or dassie, Waterberg National Park, Namibia. Although rodentlike in appearance, hyraxes have more in common with hoofed ungulates such as the elephant.

THE HYRAXES ARE SOMETHING of a zoological puzzle. They were first grouped with rodents due to their appearance, and later with the rhinoceroses and elephants because they are ungulates. Today the five species of rock hyraxes (*Procavia*), the three species of gray, bush or yellow-spotted hyraxes (*Heterohyrax*) and the three species of tree hyraxes (*Dendrohyrax*) are placed in a single family in the order Hyracoidea.

Hyraxes are small, thickset, grayish brown and rabbitlike. They are tailless or have only short tails, ⅖–1⅕ inches (1–3 cm) long, and have small round ears and short muzzles. The largest species is no more than 2 feet (60 cm) in length. Their forefeet have four functional toes, the fifth being a mere stump. The hind feet have three toes, the inner toe having a curved claw. All the other toes have short, hooflike nails.

The best-known species are the Cape hyrax or dassie of Cape Province and southwestern Africa and the daman or cherogil of Syria to Sinai and Arabia. There are many species and subspecies ranging across the Transvaal, Angola, East Africa, Ethiopia, Sudan, the Democratic Republic of Congo (Zaire), Nigeria and Cameroon. Rock hyraxes usually live in the drier, rocky areas of Africa and Arabia, while gray or yellow-spotted hyraxes are mainly found in the eastern half of Africa. Tree hyraxes live in forests in equatorial Africa scattered down the east coast, in Bioko off Cameroon and in Zanzibar.

Zoological curiosities

When the Bible was translated into English, *coney*, meaning a rabbit, was the nearest the translators could get to naming this strange type of mammal. The name hyrax is Greek for shrew, but the first scientific account of the animal was not given until 1766, when it was called *Cavia capensis*, a kind of guinea pig. The first Dutch settlers in South Africa called it *dasje* or little badger, now spelled *dassie*.

A hyrax is an unusual mixture with its rodentlike appearance, but many characteristics of a hoofed animal. Its skeleton has much in common with that of the rhinoceros, but it has a larger number of ribs. Its teeth are probably its most unusual feature, with some like those of the rhinoceros and others like those of the hippopotamus. Since teeth are much used in classifying mammals, this has been the source of classification problems in the past.

The bones of a hyrax's forelegs and feet are like those of an elephant. Its brain is also similar to that of an elephant, but its stomach is more like that of a horse. Its hind feet, with the three toes and hooflike nails, recall the hind feet of the horse's ancestor. To complete this mixture, the hyrax has a gland in the middle of its back surrounded by lighter hairs, which are erected in moments of excitement. A gland like this is found on the top of the head of the capybara, *Hydrochaeris hydrochaeris*, the South American rodent, which also has hind legs with three toes.

Noisy alarm calls

Rock hyraxes are usually timid creatures, but they can be aggressive at times. Their safety lies in being able to seek shelter quickly. When feeding, they have a lookout that gives warning of danger by shrill shrieks. Hyraxes have a loud call, and this alone probably gives them protection, for danger or "all clear" signals can be heard over a wide area. Tree hyraxes seek safety in holes in trees. Hyraxes have neither sharp claws nor grasping fingers, but they do have rubbery pads on their feet. They use these to scamper over smooth faces of rock or, in the case of tree hyraxes, to run up smooth trunks.

Although all hyraxes are active mainly by day, they are on the alert on moonlit nights, when their calls may be heard. The call is a mewing note that may rise higher and higher to end in a prolonged scream. This might be answered by another hyrax as far as 1 mile (1.6 km) away. The alarm note, as when a bird of prey flies overhead, is a short, coarse bark.

HYRAXES

CLASS	**Mammalia**
ORDER	**Hyracoidea**
FAMILY	**Procaviidae**
GENUS	**Rock hyraxes, *Procavia*; bush, gray or yellow-spotted hyraxes, *Heterohyrax*; tree hyraxes, *Dendrohyrax***
SPECIES	***Procavia capensis; Heterohyrax antinea; H. brucei; H. chapini; Dendrohyrax dorsalis; D. arboreus; D. validus*; others**

ALTERNATIVE NAMES
Dassie; coney

WEIGHT
About 8⅘ lb. (4 kg)

LENGTH
Head and body: 1–2 ft. (30–60 cm)

DISTINCTIVE FEATURES
Compact body; blunt snout with cleft upper lip; short ears; short, sturdy legs; very short tail; grayish brown fur

DIET
Grasses, herbs and leaves

BREEDING
Age at first breeding: 16 months; breeding season: varies geographically; gestation period: 202–245 days; number of young: 1 to 6; breeding interval: 1 year

LIFE SPAN
Up to 15 years

HABITAT
Rock hyraxes and gray or yellow-spotted hyraxes: rocky areas and open grassland. Tree hyraxes: forest and woodland.

DISTRIBUTION
Most of Africa; western Middle East

STATUS
Generally common; some species vulnerable

Hyraxes

Family groups

Colonies of hyraxes may contain between 6 and 50 individual animals. The larger colonies are made up of family groups of females and their young and one older male. When alarmed, the hyraxes scamper for a hole in the rocks or in the ground, the male bringing up the rear.

The daily pattern is for all members of the colony to come out at dawn to sun themselves on a rock, all grouped together. As the sun warms up, they slowly spread out, grooming and stretching. Then they feed for about an hour, tree hyraxes climbing trees to feed on leaves, fruits and twigs, rock hyraxes feeding mainly on grass but also on herbs and low bushes. Gray or yellow-spotted hyraxes also feed on leaves, especially acacia. An hour before noon they move into shade to rest and come out feed again later as the sun goes down.

Slow breeding, long life

There are usually two or three young, although there may be up to six in a litter, born after a very long gestation of 202–245 days. The babies are born fully furred and with their eyes open, and are able to begin nibbling food after a few days. Hyraxes become sexually mature at 16 months. Their maximum life span is 15 years, which is long for a mammal of their size.

Slow breeders and with a large number of predators, some species of hyraxes are now classified as vulnerable. These include two species of gray or yellow-spotted hyraxes and all species of tree hyraxes. Snakes such as cobras and puff adders, large eagles and Mackinder's owl are their main predators. They are also taken by leopards, Cape hunting dogs or African wild dogs, caracals and mongooses.

Hyraxes live in large family groups called colonies. Feeding, sunbathing and defending the territory are all group activities.

IBEX

The Nubian ibex, a subspecies of the Alpine ibex, used to be widespread, but is now endangered. An adult male is pictured.

THE NAME IBEX IS APPLIED to several sturdy, sure-footed high-mountain forms of wild goats. They are found on rocky mountain slopes and montane pastures in Europe, Asia and northeast Africa, normally at or above the tree line. Ibexes are a varied group and it is still not agreed whether or not they are all closely interrelated. It is thought that some species may have been derived independently from the true wild goat or bezoar, *Capra aegagrus*, which inhabits low-lying but hilly desert country. Most ibexes are larger and more thickset than the wild goat, however, and their horns are broad and rounded in front, with evenly spaced knots on the surface. This compares to the wild goat's horns, which tend to be narrow and keeled on the front surface.

The horns of most ibexes are scimitar-shaped or curved, but the Spanish ibex (*Capra pyrenaica*) and the East Caucasus ibex (*C. cylindricornis*), locally known as the tur, have very divergent, twisted horns. One well-studied subspecies related to the Alpine ibex (*C. ibex*) is the Siberian ibex (*C. ibex sibirica*). This subspecies has the longest horns. The Caucasian turs have thick horns, whereas in the wild goat and another subspecies, the Nubian ibex (*C. i. nubiana*), the horns are slender. In the Siberian ibex and the walia ibex (*C. walie*) the knots on the front surface of the horns are bold and well developed. In the Nubian, Alpine and West Caucasus ibexes, however, the knots are clear only toward the inner edge. Females of all forms are similar, with short, back-turned horns.

Change coats in winter

In summer most ibexes are chocolate brown in color. Some forms have black stripes down the outside of the limbs, along the lower flanks, along the middle of the back and down the nose, while the belly is white. These markings are sometimes almost or entirely lost, and are always less developed in the females. In winter the coat changes to become far thicker and the coloring becomes more variable with species. For example, in some Siberian ibex there is a large white "saddle" mark in winter.

Ibexes can weigh up to 330 pounds (150 kg), with females weighing rather less than males. The head and body length is usually 3⅗–5½ feet (1.1–1.7 m) and they stand around 2–3⅖ feet (65–1 m) at the shoulder.

Up and down the mountains

Thanks to the work of Russian zoologists, the habits of the Siberian ibex are fairly well known. It lives at 1,700–17,000 feet (520–5,200 m) above sea level, even higher in the Pamir, in steeper country than the wild sheep living in the same mountains. Herds consist of 3 to 40 animals, rising to 200 in places, especially during the coldest months. In winter the ibexes move to steeper slopes with less snow, often south-facing. This may involve a downward migration of 1,000–6,500 feet (300–2,000 m), which sometimes brings the ibex near to the forest zone, where some stay the whole year. At night ibex may move down the mountainside to avoid frosts and move up again to feed later in the morning. They mainly graze on grass and herbs in the summer, but browse on leaves in the winter months.

Enforced celibacy

The rut (mating) takes place in autumn or as late as December and early January in some parts of Europe and Central Asia. It lasts between 7 and 10 days, the males feeding little at this time.

IBEXES

CLASS **Mammalia**

ORDER **Artiodactyla**

FAMILY **Bovidae**

GENUS *Capra*

SPECIES Alpine ibex, *Capra ibex*; Spanish ibex, *C. pyrenaica*; Walia ibex, *C. walie*; West Caucasus tur, *C. caucasica*; East Caucasus tur, *C. cylindricornis*

ALTERNATIVE NAMES
Siberian ibex (*C. ibex sibirica*); Nubian ibex (*C. i. nubiana*)

WEIGHT
77–330 lb. (35–150 kg)

LENGTH
Head and body: 3⅗–5½ ft. (1.1–1.7 m)

DISTINCTIVE FEATURES
Heavy body; short, sturdy legs; long, curved horns; thick beard; chocolate brown with pale belly (summer); variable color (winter)

DIET
Grasses and herbs; also leaves in winter

BREEDING
Age at first breeding: 3–6 years; breeding season: autumn or winter; gestation period: 150–180 days; number of young: usually 1; breeding interval: 1 or 2 years

LIFE SPAN
Up to 22 years in captivity

HABITAT
Rocky mountain slopes and montane pastures

DISTRIBUTION
Spain, European Alps, northeastern Africa, Middle East and parts of south-central Asia

STATUS
Endangered: Spanish ibex, Nubian ibex, Walia ibex and West Caucasus tur; vulnerable: East Caucasus tur; locally common: Alpine ibex and Siberian ibex

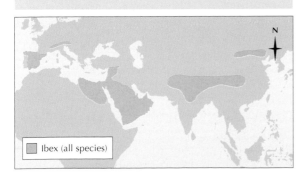

Ibex (all species)

They fight among themselves to form harems of females, rearing on their hind legs and clashing their horn-bases together. This specialized method of fighting causes no injury but ensures that the largest males with the thickest horns win. As a result, a male has not much chance of gathering a harem until around 6 years old, although he may become sexually mature at 1½ years. The older males are able to gather large harems in which each year 15–20 percent of the females stay barren.

A herd of Alpine ibex, Switzerland. Ibex live on rocky mountain slopes, grazing in summer. In winter they move down near the tree line where they browse on leaves.

Single young

Gestation lasts 150–180 days, and the single young, weighing 8–9 pounds (3.5–4 kg), is born in around May. Only about 5 percent of females have twins. The young are not weaned until the autumn but begin to graze after a month. Siberian ibex are thought to live 15–20 years in the wild, while captive ibex have been known to live for 22 years. In the Nubian ibex the rut occurs in September and October. Gestation is shorter, so the young are born from February to April.

The young ibexes often hide by day in the rocks, but are still preyed on by eagles, jackals and foxes. Predators of adult ibexes include leopards, snow leopards, bears, lynxes and wolves. Ibexes normally seek safety in flight but can fight with their horns when cornered. They have also been heavily hunted in the past, and the Nubian and Spanish ibexes are now extremely rare.

IBIS

I BISES BELONG TO THE ORDER Ciconiiformes, together with the egrets, herons, bitterns and flamingos. They have similar long, spindly legs, long necks and long bills. Their necks and bills are, however, generally stouter than those of herons and their bills are down-curved. Ibises also lack the powder-down patches of the herons.

The smallest ibis is the glossy ibis, *Plegadis falcinellus*, which is about the size of a curlew, with a dark plumage that shines with iridescent greens and purples. It is wide-ranging, being found in the southeastern United States, the Caribbean, southern Europe, Asia (including Indonesia), Australia, parts of Africa and Madagascar. The sacred ibis, *Threskiornis aethiopicus*, is white with a black head and neck and a black "bustle" of feathers over the tail as in the cranes. The dark wing tips are prominent in flight. It lives in Africa south of the Sahara, Madagascar and Arabia. The scarlet ibis, *Eudocimus ruber*, lives in tropical America from Venezuela to Brazil, while the white ibis, *E. albus*, is wholly white except for red on the bill and the

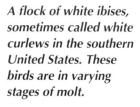

A flock of white ibises, sometimes called white curlews in the southern United States. These birds are in varying stages of molt.

naked skin on its face. It ranges from the southern United States, where it is often called the white curlew, to northern South America. It is sometimes thought that the white and scarlet ibises might be a color variant of the same species.

Two species under threat

These are the best known of the 25 or so species of ibises, most of which remain fairly common. However, one that has become very rare is the Japanese crested ibis, *Nipponia nippon*. In 1998 only around 50 birds were known, all from Shaanxi Province in China. The other species that is under threat is the northern bald ibis, hermit ibis or waldrapp, *Geronticus eremita*. It used to breed in central Europe but has been extinct there for three centuries. By 1994 numbers in western Morocco had fallen to just 60 pairs in 4 colonies, with perhaps 200 individuals in all. In 1996, 38 birds were found dead. However, some northern bald ibises have been recently recorded passing through Yemen and Saudi Arabia.

SCARLET IBIS

CLASS	**Aves**
ORDER	**Ciconiiformes**
FAMILY	**Threskiornithidae**
GENUS AND SPECIES	***Eudocimus ruber***

WEIGHT
18–27 oz. (510–770 g)

LENGTH
**Head to tail: about 23 in. (58 cm);
wingspan: about 33 in. (85 cm)**

DISTINCTIVE FEATURES
**Extremely long, down-curved bill; long
neck; long, spindly legs; bright scarlet
plumage with black wing tips**

DIET
**Mainly crustaceans, especially fiddler
crabs; also bivalve mollusks, aquatic
insects and small fish**

BREEDING
**Age at first breeding: 2 years; breeding
season: usually April–August, eggs laid
mainly during rainy season; number of
eggs: usually 2; incubation period: 21–23
days; fledging period: 35–42 days; breeding
interval: probably 1 year**

LIFE SPAN
Not known

HABITAT
**Mangrove swamps, muddy estuaries and
freshwater marshes; nests on islands with
mangrove trees**

DISTRIBUTION
**Venezuela south to northernmost Brazil
and northern Ecuador; also on island of
Trinidad, off Venezuela**

STATUS
Locally common

Scarlet ibis

Picturesque birds

Ibises, along with other long-legged and long-necked birds such as herons, cranes, flamingos and storks, are among the most attractive of birds. Their plumage, for example that of the scarlet ibis, is often extremely striking, especially when they are seen in huge flocks. Ibises also have a statuesque poise and rhythmic, graceful flight. They fly with their necks and legs extended, alternately flapping their wings and gliding. In flight a group of ibises will often beat their wings in unison and then sail, the front of the flock stopping first and those behind following suit, so ripples of gliding and wing-beating pass down the column.

Sacred chimney sweep

Ibises are wading birds and usually live in marshes, shallow lakes and along the shore. Here they use their slim bills to probe for small water animals, such as crustaceans, frogs, fish, worms, mollusks and small reptiles. Some species are more terrestrial, the northern bald ibis, for example, being found in the much drier country of North Africa and the Middle East. Here it feeds on beetles, grasshoppers and other small land animals. Meanwhile in South Africa the sacred ibis is called the "chimney sweep" because it eats carrion and "sweeps" out the insides of carcasses for the insects feeding there. Feeding sites may be located some distance from the roost.

The scarlet ibis has striking red plumage. Like other ibises, it uses its long, down-curved bill to probe for insects, crustaceans and small fish in the shallow water of swamps and marshes.

Scarlet ibises in flight, Canelas Island, off the coast of Para State, Brazil. This species is known for its large nesting colonies, although flocks in the thousands are not as common as they were.

Trees essential for nesting

Ibises build nests in trees except for the northern bald ibis, which nests on cliffs, and the sacred ibis, which sometimes nests on the ground. The Japanese crested ibis nests in tall forest trees. One reason for the decline of this species is the cutting down of forests. Even if the nesting tree is left intact while the surrounding trees are cut down, the ibis will desert it. This species is unusual in that it nests in pine trees. The nests are untidy platforms of sticks and rushes.

The black or red-naped ibis, *Pseudibis papillosa*, of India, sometimes uses the abandoned nests of birds of prey. This ibis nests in groups of two or three pairs. Other species are more likely to nest in colonies, although these vary considerably in size. The scarlet ibis, for example, nests in colonies sometimes thousands strong. These colonies are often shared with other birds such as cormorants, herons and egrets. Such concentrations have declined in recent years and are now much rarer as the scarlet ibis has been slaughtered for its bright-colored feathers as well as for its meat. Nonetheless, it is still fairly common across most of its range, as are most other species.

It is thought that most ibises form pair bonds that last the season. Both parents incubate the three or four white or blue eggs, for about 21 days. When the parents swap places at the nest, they will often bill and coo, preening each other and calling quietly. The chicks are also fed by both parents, with the adults regurgitating food directly to the young. When a couple of weeks old, the young leave the nest and climb out onto the nearby branches. They have two coats of down before growing their dark, immature plumage and are capable of flight between 30 and 50 days. Adult plumage is attained after between 1 and 2 years.

The sacred ibis

At one time the sacred ibis bred along the banks of the Nile, and the ancient Egyptians held it in great esteem, identifying it with the god Thoth, who was said to record the life of every man. Thoth is shown in pictures with the head of an ibis, and ibises were tamed and kept as pets in temples. They were also mummified and buried in the tombs of Pharaohs, perhaps to record their final voyage to the next world.

ICE FISH

THE FIRST REPORTS OF ICE FISH were probably those of Norwegian whalers working in the Antarctic, who brought back stories of "bloodless" fish they caught near their shore factories. Ice fish do, in fact, have blood, but it is almost transparent, with a just perceptible yellowish tint. It lacks hemoglobin, the red pigment that in many other animals carries oxygen from the lungs or gills to other parts of the body. Scientists have long debated how ice fish survive without the oxygen-carrying capacity of hemoglobin, but due to the remoteness of the Antarctic they could not solve the problem until relatively recent times.

Ice fish should not be confused with the unrelated icefish, a group of small semitransparent fish in the family Salangidae. The latter live in the seas off Korea, China, Japan and Siberia.

Translucent, scaleless fish

The name ice fish was given by British whalers in allusion to the translucent appearance of the body. Ice fish have no scales, and the body is very pale brown or whitish, and slimy. The front of the head is drawn out into a beak with a large, gaping mouth edged with thick lips. The eyes are large and goggling. The dorsal fin is divided into two parts; the front part in the middle of the back is sail-like and the second part is ribbonlike, similar to the anal fin on the underside. The large pectoral fins form paddles just behind the gills, and the fleshy leglike pelvic fins lie in front of them under the belly.

Confined to icy seas

There are 17 species of ice fish, all but one of them confined to the cold Southern Ocean. The exception, the pike ice fish, *Champsocephalus esox*, ranges north as far as Patagonia and the Falkland Islands. The largest species, the blackfin ice fish, *Chaenocephalus aceratus*, measures up to 30 inches (75 cm) long and can weigh 2½ pounds (1.1 kg).

Ice fish tend to occur in coastal waters, often where local upwellings increase the food supply. The ocellated ice fish, *Chionodraco rastrospinosus*, is found at depths of 660–1,300 feet (200–400 m), whereas *Chaenodraco wilsoni* lives in shallower waters at depths of less than 800 feet (245 m).

Ice fish such as this blackfin ice fish live in the bitterly cold seas of the Antarctic. They are unique among vertebrates because their blood lacks hemoglobin and contains antifreeze.

Sluggish predators

The fish of the Antarctic are now being studied intensively by scientists of several nations, but by comparison with other species, such as the Antarctic cods of the family Nototheniidae, little is known of the habits of ice fish. They have been caught with nets at depths of about 2,100 feet (640 m) in some numbers.

The muscles of ice fish are weak and their ribs soft, which suggest that the fish are not active. They probably spend much of their time on the seabed resting on their leglike pelvic fins, engulfing passing fish or picking up carrion. They probably have big meals at long intervals. An ice fish's large mouth can close over a fair-sized Antarctic cod, and the stomach and skin of its belly can stretch to accommodate a large meal. The proof of this is that ice fish are sometimes caught when they have engulfed an Antarctic cod already hooked. Ice fish also catch krill, the shrimplike crustaceans that abound in the cold, oxygen-rich waters, supporting whales, penguins and many other Antarctic animals.

Breeding

Most species of ice fish spawn in the Antarctic autumn, between mid-March and late April. The pike ice fish spawns from February to September. Females lay up to 2,000 large yolky eggs, about 4 millimeters in diameter, on the sea floor.

Blood is unique among vertebrates

The discovery that ice fish have no hemoglobin in their blood posed several questions. The first was how they manage to transport oxygen to their tissues, for in red blood 90 percent of the oxygen taken into the body is carried by the hemoglobin and the rest is dissolved in the blood plasma. Ice fish must carry all their oxygen in the plasma. They are helped by the high concentration of oxygen in the Antarctic seas. Gases dissolve in cold water better than in warm water. As a result, the organisms living in the cold seas, where temperatures rarely rise more than a few degrees above freezing point, have a greater supply of oxygen and they oxygenate their bodies more efficiently. Ice fish are probably able to absorb oxygen through the scaleless skin as well as through the gills.

Ice fish also have large hearts, about three times the size of the hearts of red-blooded fish. This must enable them to pump blood rapidly through the body and so compensate for their lack of hemoglobin. Perhaps the most extreme example in ice fish of adaptation to life in cold seas is the presence of natural antifreeze in their blood. The blood contains AFP (antifreeze glycopeptides), which enable body fluids to remain liquid down to 28° F (-2° C).

ICE FISH

CLASS	**Osteichthyes**
ORDER	**Perciformes**
FAMILY	**Channichthyidae**
GENUS	***Chaenocephalus, Champsocephalus, Channichthys, Chaenodraco, Cryodraco, Chionodraco, Chionobathyscus, Pagetopsis, Neopagetopsis, Pseudochaenichthys***
SPECIES	**17, including blackfin ice fish, *Chaenocephalus aceratus* and pike ice fish, *Champsocephalus esox***

WEIGHT
***Chaenocephalus aceratus*: up to 2½ lb. (1.1 kg); other species lighter**

LENGTH
***C. aceratus*: up to 30 in. (75 cm); other species shorter**

DISTINCTIVE FEATURES
Beaklike head; large jaws with thick lips; large, prominent eyes; dorsal fin divided into sail-like front section and elongated, ribbonlike back section; body pale brown and almost translucent; scales absent

DIET
Krill; fish such as Antarctic cod

BREEDING
Breeding season: February–September (*Champsocephalus esox*), March–April (other species); number of eggs: up to 2,000

LIFE SPAN
Not known

HABITAT
Cold seas, often near local upwellings

DISTRIBUTION
***Champsocephalus esox*: waters off southern South America (Patagonia and Falkland Islands). Other species: Antarctic seas.**

STATUS
No species threatened

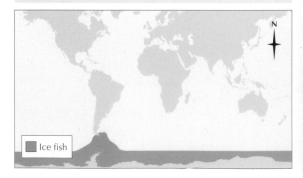

■ Ice fish

ICHNEUMON WASP

ICHNEUMONS AND BRACONIDS are small, parasitic wasps, belonging to the superfamily Ichneumonoidea. They are often referred to as parasitoids because although their larvae lead a parasite-like existence in the bodies of other insects and spiders, there is some debate as to whether they qualify as true parasites. There are other parasitic species in the same order, Hymenoptera, including many of the chalcid wasps, but only the ichneumons and the braconids are discussed here.

A living death

Female ichneumon wasps and braconids have an unusual method of ensuring that the larvae hatching from their eggs have a plentiful supply of fresh food throughout the larval stage.

The ichneumon wasp, for example, first chooses a host insect, such as the caterpillar of a moth or butterfly. She makes sure, in some way scientists do not yet fully understand, that the host has not already been visited by another parasitoid. The female then lays an egg inside the caterpillar using a slender ovipositor (the egg-laying organ) that arises from the abdomen. The caterpillar might rear up and thrash about when the wasp settles on it, but once the egg has been laid the host resumes its feeding and growing in an apparently normal way. Meanwhile the egg of the ichneumon wasp hatches into a tiny grub, which then begins feeding and growing at the expense of the caterpillar's tissues.

When the time comes for the caterpillar to pupate, it is harboring in its body a parasite of fair size, but not large enough to destroy the working of any of its essential organs. As soon as the change to a chrysalis is completed, however, the grub begins to grow apace, soon killing its host. It then turns into a pupa itself, within the empty shell that should contain a moth or a butterfly. Some time later the adult ichneumon wasp hatches by splitting the skin of its own pupal covering, then chewing its way out through the shell of the chrysalis. A moth or butterfly pupa that has harbored an ichneumon wasp or a braconid is never split open as it is when it completes its normal transformation, instead it has a hole in the side.

If the hatching ichneumon wasp or braconid is a female, she will find a male of her own species and mate. Then, if summer is nearly over, the female will probably overwinter (hibernate through the winter). The following spring, having emerged from hibernation, she seeks out another caterpillar, possibly but not necessarily of the same species that provided her with a living food supply. The entire life cycle is then repeated.

No escaping the parasitoids

The very common, large, rusty-red ichneumon wasps of the genus *Ophion* have this sort of life history, but variations on the reproductive pattern are numerous. Not all ichneumons parasitize the larvae of moths and butterflies. Some attack sawflies, while others parasitize beetles, bugs or spiders. Aphids are heavily attacked by small braconids of the genus *Aphidius*. One species even goes underwater to lay its eggs in the larvae of caddis flies.

The stage at which the parasites emerge from the host's body may vary. It may be before pupation or after. The stage at which the eggs are laid may also vary. In the true ichneumons it is usually in the host's larva.

One of the largest ichneumon wasps, *Rhyssa persuasoria*, lays its eggs inside the wood-boring larvae of horntails (genus *Sirex*), which are large sawflies. The horntail larvae burrow in pine trees, ruining the timber. The female ichneumon has a very long, slender ovipositor and is able to bore through a couple of inches of solid wood and implant an egg on a horntail grub inside the trunk. The drilling through the wood is a remarkable performance, but exactly locating the grub within is even more so, the ichneumon using its antennae to "sniff out" the larva.

Multiple parasites

Some parasitoids lay not one, but a number of eggs, resulting in a brood of 100 or more grubs inside one host. This multiple parasitization is

Female ichneumon wasps have long, slender ovipositors with which they lay their eggs in their chosen hosts. Pictured is **Rhyssa persuasoria.**

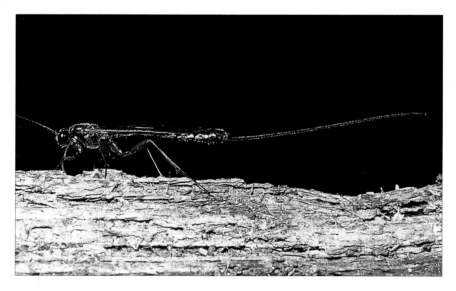

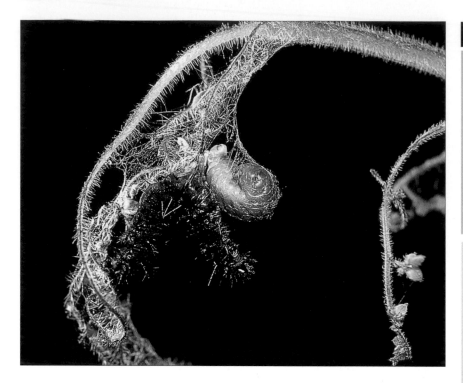

Ichneumon and braconid larvae eat their hosts alive before pupating and turning into adult insects.

ICHNEUMON WASPS

PHYLUM	**Arthropoda**
CLASS	**Insecta**
ORDER	**Hymenoptera**
SUBORDER	**Apocrita**
SUPERFAMILY	**Ichneumonoidea**
FAMILY	**Ichneumonidae and Braconidae**
SPECIES	**About 40,000**

ALTERNATIVE NAME
Ichneumon fly

LENGTH
Largest species (*Megarhyssa* spp.): up to 2 in. (5 cm); most species: 1–2 mm

DISTINCTIVE FEATURES
Resemble small wasps, with elongated, slender, curved abdomen; longer antennae than typical wasps; female's ovipositor (egg-laying tube) may be longer than body

DIET
Adult: flower nectar. Larva: juices of host insect, usually a butterfly or moth larva (caterpillar); some species parasitize spiders.

BREEDING
Number of eggs: 1 to 100, laid inside host; larval period: several weeks to several months, some species overwinter as larvae

LIFE SPAN
Several weeks to 1 year

HABITAT
Varies according to habitat of host species

DISTRIBUTION
Virtually worldwide

STATUS
Most species common

characteristic of the Braconidae, most of which are very small insects. One of these, *Apanteles glomeratus*, is a serious and, from the human point of view, useful, parasite on cabbage white butterflies. In this case the larvae of the braconids emerge from the body of the caterpillar just at the time when it has found a place to pupate. On emergence from the caterpillar, each braconid larva spins a small, yellow cocoon, like a tiny replica of that of a silkworm. These clusters of cocoons are common on walls in vegetable gardens.

Parasites on parasites

The cocoons contain quantities not only of the *Apanteles* braconid, but often numbers of another, even smaller parasite. The females of this other species are expert at finding butterfly caterpillars infested with *Apanteles* larvae and lay eggs on the first parasites, probing through the caterpillar's skin to do so. The term *hyperparasite* is used to describe this sort of parasite within a parasite.

Pest control

Braconids are extremely important in controlling aphids, and both ichneumon and chalcid wasps have been used in the biological control of harmful insects. In New Zealand, horntails were accidentally introduced into imported timber and bred rapidly, becoming a serious pest in the pine forests and plantations. There is no parasitoid native to New Zealand that can reach the horntail larva in its burrow, so in 1928 and 1929 over a thousand pupae of the braconid *Rhyssa persuasoria*, were sent to New Zealand. The braconids were released in the pine woods, and horntails are now far less abundant there.

Parasites or not?

A truly parasitic animal does not kill the host on which it feeds. On the contrary, it is in its interest for the host to stay alive as long as possible, since if it dies the parasite perishes with it. The case of the ichneumon larva is quite different. Here the host is doomed to die from the moment the egg is deposited in it. Ichneumon wasps prey on caterpillars, their larvae slowly eating them alive instead of killing them immediately like a normal predator. For this reason some authorities object to the term parasite to describe ichneumons and braconids and instead call them parasitoids.

IGUANA

IGUANAS ARE LARGE, often herbivorous lizards. They occur in the Americas, on the Galapagos Islands, on many islands in the Caribbean and, curiously, also in Madagascar and Fiji. Similar large lizards in Australia are often called goannas, a corruption of the word iguana, but these are predatory and do not belong to the family Iguanidae. Here we mainly consider the green or common iguana (*Iguana iguana*) of Central America and northern South America, but also look at the terrestrial or land iguanas (genus *Conolophus*) and the rhinoceros iguanas (genus *Cyclura*). There is also the much smaller desert iguana, *Dipsosaurus dorsalis*, which ranges from Central America northward into California, Arizona and New Mexico. The unique marine iguana of the Galapagos Islands, *Amblyrhynchus cristatus*, is described elsewhere.

"Prehistoric monsters"

Green iguanas are found in the forests and scrublands of Central America and northern South America, and spend most of their time among the branches. Since many people now keep them as pets, they are familiar in parts of North America too. As the name suggests, they are usually green in color. They also have a crest of spiny scales along the back. These are particularly well-developed in the adult males, as is the dewlap, a large pendulous flap of skin that is used in courtship and to deter rivals. Green iguanas can grow to a length of 6 feet (1.8 m) but most individuals are smaller than this. These iguanas are among the living lizards that are closest to what most people think "prehistoric monsters" would look like.

Terrestrial or land iguanas are also found in Central and South America, on many islands in the Caribbean and on several islands of the Galapagos group. The crest and dewlap are not so well developed as in green iguanas. There are a number of species, most of which are a uniform brown color. The rhinoceros iguanas have developed two or three hornlike scales on the head, giving them their name.

Smaller desert species

Desert iguanas are much smaller. Adults are rarely longer than 12 inches (30 cm) in total length. They are found in northern Mexico and Baja California and in the United States in southeastern California, western Arizona, the extreme south of Nevada and a tiny corner of Utah. They do not have dorsal spines, but they do have a row of enlarged scales along the length of the back. The base color of desert iguanas is pale gray, but most individuals have varying amounts of darker spots and blotches over the body.

The green or common iguana is an agile climber and spends most of its time in the branches of tropical forests in Central America and northern South America.

Iguanas are mainly vegetarian, feeding on a variety of plant material including leaves, shoots, flowers and fruits.

Dash to safety

Green iguanas are agile climbers and they are rarely found far from the trees of the tropical forests in which they live. They are able to move surprisingly fast and are usually very difficult to capture. They bite and their long claws can inflict nasty scratch wounds. Green iguanas particularly favor trees that overhang the water and will dive in fearlessly if they are disturbed. They are powerful swimmers.

The green iguana comes down to the ground in cold weather and hides under logs or in holes, but other iguanas are usually ground-living and only occasionally climb trees. The desert iguana is a very fast runner, using only its hind legs as it sprints upright across the ground.

Vegetarian lizards

Iguanas are primarily vegetarian. Green iguanas eat a variety of plant foods, including shoots, fruits, flowers and leaves, although juveniles of this species feed on insects at first because they contain vital protein. The desert iguana prefers the yellow-flowered creosote bush and in the United States is common in deserts dominated by this plant. It also eats other flowers, and only when the flowering season is over does it feed on insects and carrion. Land iguanas feed on cacti, although the larger species occasionally take small rodents.

IGUANAS

CLASS	**Reptilia**
ORDER	**Squamata**
SUBORDER	**Sauria**
FAMILY	**Iguanidae**

GENUS **Tree iguanas, *Iguana*; land iguanas, *Conolophus*; rhinoceros iguanas, *Cyclura*; desert iguanas, *Dipsosaurus***

GENUS AND SPECIES **Many, including green iguana, *Iguana iguana*; and desert iguana, *Dipsosaurus dorsalis***

ALTERNATIVE NAME
Green iguana: common iguana

LENGTH
Green iguana: up to 6 ft. (1.8 m); most species smaller

DISTINCTIVE FEATURES
Most species large, with crest of spiny scales along back; dewlap (loose flap of skin on throat) in some species; background color usually green, pale gray or brown

DIET
Shoots, leaves, flowers and fruits; certain species also take insects, carrion and rodents

BREEDING
Green iguana. Number of eggs: 20 to 70; hatching period: about 90 days.

LIFE SPAN
Varies according to species

HABITAT
Green iguana: forests. Land iguanas: open scrub. Desert iguana: deserts.

DISTRIBUTION
Central and South America; Galapagos Islands; Caribbean; Madagascar; Fiji

STATUS
Green iguana: common; most other large species endangered

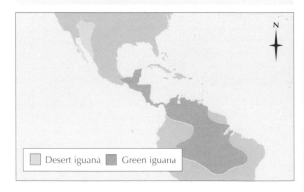

| Desert iguana | Green iguana |

Tasty flesh

Iguana flesh is tender and very tasty and local people have captured them, for both their meat and their eggs, from time immemorial. Now, in a move that makes a lot of sense, green iguanas increasingly are being farmed. Iguanas will eat a wide range of food, including most household scraps, and they grow fast in hot climates. It takes less food to produce a pound of iguana meat than a pound of pork or chicken, or of any other warm-blooded animal for that matter. This is because food energy does not need to be expended on maintaining a high body temperature. Similar arguments apply to the farming of crocodiles, which is also practiced in many parts of the tropical world. Iguanas have the advantage that they are easier to handle than crocodiles, but the leather made from the skin of crocodiles is thicker and much harder-wearing.

Young grow rapidly

Female green iguanas lay a clutch of between 20 and 70 eggs in a burrow dug especially for the purpose in the soil. The eggs hatch after about 3 months and the babies are about 10 inches (25 cm) in length. By the time they are one year old, they may be 3 feet (90 cm) in length. This is good for iguana farmers, but often comes as a nasty shock to people who have bought a baby iguana as a pet, not realizing that it will rapidly grow into a voracious and powerful animal.

Vanishing iguanas

Green iguanas are still abundant in many parts of their range, but most other species of iguanas are endangered. When Charles Darwin visited the Galapagos Islands in 1835, land iguanas were extremely abundant. He wrote "I cannot give a more forcible proof of their numbers than by stating that when we were left at James Island, we could not for some time find a spot free from their burrows to pitch our single tent."

Now, on many of the Islands, iguanas are rare. Numbers have been decimated by goats, which eat their food and remove the cover that protects them. Pigs and rats are also serious predators because they eat the eggs, while dogs and native hawks are a threat because they prey on the iguanas' young. Human beings, meanwhile, do all of these things as well as destroying the iguanas' habitat. It is the same story in many of the more populous parts of Central America and on many of the islands of the Caribbean. Desert iguanas in the United States are threatened by habitat destruction and fragmentation and they are now strictly protected.

But not all is gloom. Around at least one Air Force base in the Caribbean, land iguanas have thrived due to protection and supplemented feeding. Bored airmen have become extremely fond of these animals and woe betide anyone who interferes with "their" iguanas, which have become increasingly tame.

A land iguana on the Galapagos Islands. Historically abundant, these iguanas are now rare on the Islands. Their young and eggs are taken by domestic animals and rats, while they also suffer habitat destruction by humans.

IMPALA

Impala, pictured here in Etosha National Park, Namibia, are graceful and agile antelopes. They mainly graze on grass, but also browse leaves from trees and bushes.

Taking to cover

Impala inhabit a wide area of East Africa and southern Africa. They seem to like being near water, and they avoid open country, being more usually found where there are low trees and tall shrubs. Often they live in areas without much ground cover, in scrub and thornbush country especially. Their distribution is patchy because they do not venture much into either overgrown or open land. So, although abundant in most of the Kruger National Park, South Africa, they are absent from much of the northern end of the park.

According to how suitable it is for the impala, an area may have a density of anything from seven to over 200 animals per square mile (2.6 sq km). The usual figure is 50 to 70 impala in an area this size. Concentrations are highest in the dry season, as with most African ungulates. The beginning of the dry season also happens to be the peak time of the rut (mating). In the wet season, impala are more scattered and occupy small home ranges, but they may wander as much as 15 miles (24 km) for water.

Impala both graze and browse bushes and trees, but in most areas they eat mainly grass.

Born when the grass sprouts

The rut mainly takes place in the beginning of the dry season. The lambs are born, one to each ewe, after a gestation of 180–210 days, early in the wet season when there is most food available.

In Zimbabwe the first lambs are dropped in early December and the peak of lambing is from December 15 to January 1. Two-year-old ewes, breeding for the first time, give birth later in the season than older ones. The young impala grow rapidly and in young males the horns begin to sprout in late February. Lambs are usually weaned before the next rut, at which time they may form separate bands. In the rut nearly all ewes breed, at least 97 percent of the older ones and 85 percent of the two-year-olds.

The rut begins when the rams (males) set up their territories in late May or early June. Surplus rams attach themselves to small groups of ewes, and the yearlings form small bands by themselves. The ewes live in herds year-round. At the end of the lambing season these herds number as many as 100 impala, including young. These large herds stay together from January to May, and only a few males associate with them. Then in May they break into smaller groups, which pass through the rams' territories and mating occurs. After the rut, the ram groups re-form, but groups of mixed sex and age predominate.

THE IMPALA, SOMETIMES called by its French name, *pallah*, is one of the most graceful of the antelopes. It is also known for its leaping and dodging when alarmed or under threat from predators.

About 2½–3¼ feet (0.75–1 m) high, weighing 88–143 pounds (40–65 kg), the impala has a glossy, red brown coat with a lighter brown or tan area on the flanks and legs. It has a dark stripe on each side of its hindquarters and a sharply defined, white belly, muzzle and tail. The male impala has lyre-shaped, ribbed horns, 20–30 inches (50–75 cm) long, which make one spiral turn. The ewe (female) is hornless. The impala's neck and limbs are slender and delicate.

The impala occupies a rather isolated position in the family Bovidae. In the past there have been divided opinions on whether it was more nearly related to the gazelles or to the reedbuck. Recently it has been suggested, after a study of the skull, teeth and horn cores, that the impala is more nearly related to hartebeest and wildebeest.

IMPALA

CLASS	**Mammalia**
ORDER	**Artiodactyla**
FAMILY	**Bovidae**
GENUS AND SPECIES	***Aepyceros melampus***

ALTERNATIVE NAME
Pallah

WEIGHT
88–143 lb. (40–65 kg)

LENGTH
Head and body: 3⅓–5 ft. (1.1–1.5 m); shoulder height: 2½–3¼ ft. (0.75–1 m); tail: 10–15½ in. (25–39 cm)

DISTINCTIVE FEATURES
Glossy, red brown coat; lighter brown or tan area on flanks and legs; dark stripe on each side of hindquarters; white muzzle, belly and tail; lyre-shaped horns (male only) with ridges on front surface

DIET
Mainly grasses and herbs; also leaves

BREEDING
Age at first breeding: about 18 months; breeding season: all year in equatorial areas, with peak at start of dry season; gestation period: 180–210 days; number of young: 1; breeding interval: 1 year

LIFE SPAN
Up to 13 years

HABITAT
Open woodland, acacia savanna and sandy thornbush

DISTRIBUTION
Eastern and southern Africa

STATUS
Generally common; some subspecies vulnerable or endangered

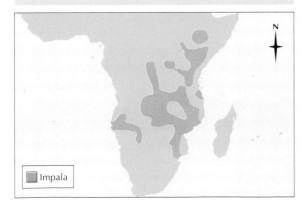

Impala

By December the groups are reduced in size to 10 or fewer. At this time the ewes become quite secretive, separating off for a while to give birth to their young.

The main predator of impala is probably the leopard. About half the young are lost to predators in the first few weeks of life. Existing populations are often subject to poaching, but this does not severely affect their numbers.

The leaping impala

Impala rams become quite aggressive in the rutting season, especially when setting up territories. At this time, fighting and chasing are common. Once the territories are set up, the rams leave their bases to drink at the waterholes, which belong to no one individual animal.

The most conspicuous piece of impala behavior, and one for which they are well known, is their alarm reaction. When disturbed, the whole group indulges in a display of leaping. They jump forward, straight up or with side turns, as much as 30 feet (9 m) into the air, up and down, around and in all directions. It has been suggested that the purpose of this behavior is to confuse a predator, such as a big cat that is trying to single out one animal from the group it is attacking. The leaping impala, helped by their contrasting colors, are often successful in preventing a predator from cutting off its prey from the rest of the herd. Such diversionary tactics might be more successful than taking flight, at least when it comes to preventing adult animals from being taken.

Impala, Zimbabwe. Leaping and jumping serve to confuse predators and may prevent a big cat from cutting off a single impala from the rest of the herd.

INDIAN BUFFALO

THE BUFFALO ARE A distinctive group of cattle. They are stockily built with large hoofs and large, shaggy ears. Their horns are triangular in cross section, instead of oval or circular like those of true cattle and bison. The head is carried horizontally, and the muzzle is broad and hairless. Buffalo have a straight back, but one that slopes down toward the hindquarters. The hair is sparse. The two genera of buffalo are *Syncerus*, the Cape buffalo of Africa, and *Bubalus*, the Asiatic buffalo.

The Asiatic buffalo differ from the Cape buffalo in the shape of their skulls, in the horns and in the forward-lying hair in the middle of their backs. The Indian buffalo or Asian water buffalo, *Bubalus bubalis*, is by far the largest of the Asiatic buffalo and is an important domestic animal. Domesticated Indian buffalo are noted for their docility, in contrast to the aggressiveness of their wild relatives. In the wild, the species reaches more than 6 feet (1.8 m) in height and weighs up to 2,645 pounds (1,200 kg). The domesticated form is smaller. The Indian buffalo is gray black in the wild, but the domesticated animal is more varied in color. It may be gray, black or white, with white spots on the chin, throat and limbs. Both males and females have horns that are heavy at the base and semicircular, spreading out sideways and then backward, in a line with the back. The spread of the horns is up to 6½ feet (2 m), the largest of any living bovid.

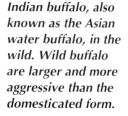

Indian buffalo, also known as the Asian water buffalo, in the wild. Wild buffalo are larger and more aggressive than the domesticated form.

Smaller forest species

There are localized populations of wild Indian buffalo in Sri Lanka, Nepal, Assam in northeast India, Indochina and Borneo.

A smaller species is the tamarau, *Bubalus mindorensis*, which is restricted to the island of Mindoro in the Philippines. It is only 3½ feet (1.05 m) high with short, thick horns that turn mainly backward and are only slightly semicircular. It is jet black in color with a few white spots, very bull-necked and weighs 600–700 pounds (272–318 kg). In historic times, the tamarau occurred on Luzon, another island in the Philippines, as well as on Mindoro.

The other two Asiatic buffalo, known as anoas, are found only on the island of Sulawesi, Indonesia. They are small and rather antelope-like with short, conical backward-pointing horns and more slender necks. The lowland anoa, *B. depressicornis*, which is the larger of the two, stands about 3½ feet (1.05 m) at the shoulder. It is black with white spots on the jaw, throat and legs. The mountain anoa, *B. quarlesi*, is about 2–3 feet (60–90 cm) in height, with a shorter tail and horns. It is golden to dark brown in color with a few white spots above the hoofs. The hair of both the anoas is long, soft and woolly. Although they have always been treated as separate species, a few scientists now think that the two anoas are variants of the same species, and that the tamarau is a dwarfed island form of the Indian buffalo.

Shy but pugnacious

Wild buffalo have a reputation for being extremely unapproachable, so they have not been studied in great detail. They are thought to live in herds of between 10 and 20 animals and are found in swampy regions, grass jungle and densely vegetated river valleys. Indian buffalo spend their time feeding on the lush grasses growing beside rivers and lakes. Classified as being endangered, there are now thought to be perhaps only 1,000 to 1,500 wild buffalo remaining in India.

The tamarau is more solitary but has been seen associating in groups of up to 11 individuals. It inhabits more rugged country, on the forest borders and in bamboo. It is even more scarce than the Indian buffalo because the forest and bamboo is cut down. Fewer than 200 or

INDIAN BUFFALO

CLASS **Mammalia**

ORDER **Artiodactyla**

FAMILY **Bovidae**

GENUS AND SPECIES ***Bubalus bubalis***

ALTERNATIVE NAME
Asian water buffalo

WEIGHT
1,540–2,645 lb. (700–1,200 kg)

LENGTH
**Head and body: 8–10 ft. (2.4–3 m);
shoulder height: 5–6¼ ft. (1.5–1.9 m);
tail: 2–3¼ ft. (0.6–1 m)**

DISTINCTIVE FEATURES
**Large size; horns heavy at base in both
sexes, usually curving backward and
inward; dark gray to black (wild buffalo);
black, gray or white (domesticated buffalo)**

DIET
Grasses growing in or beside rivers and lakes

BREEDING
**Age at first breeding: 18 months or more;
breeding season: varies according to location;
gestation period: 300–340 days; number of
young: 1; breeding interval: 2 years**

LIFE SPAN
**Wild buffalo: up to 25 years.
Domesticated buffalo: up to 29 years.**

HABITAT
**Wet grasslands, swamps, densely vegetated
river valleys and grass jungle**

DISTRIBUTION
**Native range: Nepal, Assam (India), Sri
Lanka, Indochina and Borneo. Introduced
range: mainly Australia; also Italy.**

STATUS
**Endangered; estimated wild population in
India: 1,000 to 1,500**

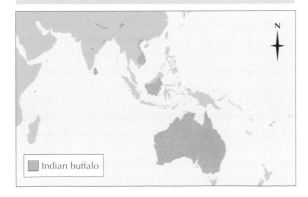

Indian buffalo

so individuals now survive in the wild. In the past tamarau could be seen grazing in the open in morning and evening, but today they have become more nocturnal.

Cattle harems

In the wild, the bulls of the Indian buffalo round up the cows into harems. At this time the males can be particularly aggressive. Both the wild and the domesticated animals may breed year-round, but in the wild animal there is a breeding peak that occurs at different times in different geographical regions. In Italy, for example, where there are semiwild Indian buffalo, the young are born between February and April. In Asia, on the other hand, most of the young are born between October and December. One calf is born after a gestation period of 300–340 days.

Buffalo become sexually mature at around 18 months, and in Italy they often breed at this age. In parts of the Far East, such as Cambodia, the domesticated animals may not mature until they are 5 years old, either because they are prevented from doing so or because their often poor diet delays maturity. Physical maturity is not reached until 3½ to 4 years.

Providing milk

Domestic Indian buffalo, bred at least partly for milk, have a long, overdeveloped lactation, like that of domestic cattle. In India and Pakistan they provide 40–50 percent of the countries' milk. In these countries, people tend to buy a female buffalo for a small sum, keep her for one or two lactations and then get rid of her because they cannot afford to keep a herd. Meat is forbidden to Hindus but, unlike the humped cattle or zebu,

Wild Indian buffalo live in herds of 10 to 20 animals. There are thought to be only 1,000 to 1,500 wild buffalo remaining in India today.

Water buffalo in paddy fields, Langkawi, Malaysia. The type known as the swamp buffalo is particularly well suited to working marshy land in humid climates. The river buffalo is better suited to producing milk.

buffalo are not in themselves sacred. Males used for hauling are often castrated, and commonly the tail is cut off as well. About half the older animals provide meat for non-Hindus.

Habitat loss the main threat

Wild Indian buffalo may occasionally be preyed on by tigers, but in general they have no serious predators except for human beings. Habitat loss is the major threat to their survival. Anoas live on an island quite free of carnivorous mammals that might prey on them, but they have, nevertheless, a reputation for being extremely aggressive. In captivity they cannot be kept with any other large animals as they will stab them in the belly with their short horns.

Swamp and river buffalo

The domestic buffalo, usually known as the water buffalo, is an important and valuable asset to its human owners. There are two main types of water buffalo: swamp buffalo and river buffalo. Swamp buffalo can work in marshy land and humid jungle. They are stocky, heavy, very strong creatures but have the drawback that they suffer in hot sun and must bathe or wallow regularly. They are used on a large scale as draft animals in the rice fields of Malaysia, Indonesia, Indochina, southern China and the Philippines, where they are called *kerabau* or *carabao*. River buffalo are more specialized, preferring dry pastures and clear rivers and canals. Used especially in India, Pakistan, southwest Asia and Egypt, they are very

docile and are kept mainly for dairy production. They are also used for meat and draft work. However, river buffalo are not used in the fields to the same degree as the swamp buffalo, which is particularly suited to work on land that has been waterlogged and in humid climates.

Widespread buffalo

There are about 80 million domestic buffalo in the world. Over half of these are in India, with about 10 million in southern China and 5 million in Pakistan. The milk-giving river buffalo of India has also been introduced into Southeast Asia and is gradually replacing the swamp buffalo, although it is not such a good draft animal. The present distribution of domesticated buffalo is of very long standing. In the Philippines, for example, they were there well before the Spaniards arrived.

Domestic buffalo are also found in smaller numbers in Europe, North Africa and western Asia. In Egypt they were unknown in the time of the Pharaohs and were probably introduced there in the 9th century C.E. via Iraq and Syria. Now there are 1.25 million Indian buffalo in Egypt. From Italy they have also been introduced into Guyana, Cayenne, Trinidad and Brazil, as well as into the lower Congo. From the island of Timor, Indonesia, around 35 Indian buffalo were imported to Melville Island, North Australia, in 1820s to serve as food for the settlers. They have since spread all over mainland Australia were they have contributed a great deal to overgrazing problems in that country.

JACANA

THESE ARE WATERBIRDS that look like long-legged coots but are in fact more closely related to shore-birds. Jacana is a Spanish word derived from the name given to the birds by South American Indians. In English the soft "c" is usually pronounced as "k." Lily-trotter, lotus-bird and water-walker are alternative names and describe the jacanas' habit of walking on floating vegetation, supported on extremely long toes. In common with coots, they have brightly colored frontal shields. On the "wrist" of the wing there is a knob or spike sometimes 1 inch (2.5 cm) long. This is said to be used in fighting.

There are eight species of jacanas living in Mexico, Central America, Africa south of the Sahara, Madagascar, Asia from India to the Malay Archipelago and eastern Australia. The northern or American jacana, *Jacana spinosa*, is found in Mexico and the Caribbean south to western Panama. It is about 9 inches (23 cm) in total length and its plumage is reddish brown with yellow-green flight feathers. It has an orange-yellow frontal shield.

The Australian lotus-bird, *Irediparra gallinacea*, which is also found in Indonesia and the Philippines, is brownish with black on its chest and the back of its neck, and white on its throat and under its tail. The white throat has an orange border, and there is a scarlet frontal shield.

Most jacanas have very short tails, but in the breeding season the pheasant-tailed jacana, *Hydrophasianus chirurgus*, of India to the Philippines, grows a long, curving tail of black feathers. The rest of its plumage is brown with white on its head and neck except for a golden patch on its neck surrounded by a band of black.

Walking on water

Jacanas are found on ponds, lakes and slow-moving rivers where there are abundant water lilies, water hyacinths and other floating plants. Outside the breeding season they may gather in flocks of hundreds or thousands of birds. The lotus-bird's range in Australia moved south when water plants invaded lagoons near Sydney.

Jacanas can run over the soft water plants because their long toes spread their weight, working in much the same way as snowshoes. Sometimes jacanas appear to be running over clear water, but they are actually being supported by perhaps two or three stems. Their gait is a

dainty high-stepping. They lift their feet high so the toes clear the surface and jerk their tails at each step. Jacanas can swim but rarely do so and, if disturbed, flutter across open water with legs and toes dangling. They feed on water plants and small animals.

Jacanas are noisy birds that make a variety of calls. Piping, churring, clucking and grunting are used by most species during the breeding season. They also make a scolding, chittering sound when alarmed or disturbed.

Like most waterfowl, jacanas breed in the later stages of the rainy season. Courtship displays consist of showing off the wings, weed carrying and bowing.

Male rears the young

Jacana nests are extremely flimsy. The three to five eggs are laid on a coil of weed or a few rush stems piled together, or even on the leaf of a water lily. The eggs are very glossy, looking as if they have been varnished, and have so many markings that the brown background is often lost. The gloss is thought to make the eggs waterproof, a useful condition, as the nest may submerge as a parent steps onto it.

The adults are very much alike, but the female is often larger than the male. In some species at least, the female plays little or no part in rearing the young and may have several mates. In the northern jacana, for example, incubation of this species' four eggs is by the male

In the bronze-winged jacana of India the female takes little or no part in the rearing of the young. She may also have several mates and may lay several batches of eggs for the males to care for.

An Australian lotus-bird, or comb-crested jacana, standing on one leg with the toes of its other leg showing from under the wings.

NORTHERN JACANA

CLASS **Aves**

ORDER **Charadriiformes**

FAMILY **Jacanidae**

GENUS AND SPECIES *Jacana spinosa*

ALTERNATIVE NAME
American jacana

WEIGHT
Male: average 2⅘ oz. (80 g); female: average 4 oz. (115 g)

LENGTH
Head to tail: 8½–9½ in. (21.5–24 cm)

DISTINCTIVE FEATURES
Chestnut or reddish brown plumage; green-glossed black head, neck and upper chest; bright greenish yellow flight feathers; orange-yellow frontal head shield; longish yellow bill; extremely long legs and toes

DIET
Insects and seeds; occasionally tiny fish

BREEDING
Age at first breeding: usually 2 years; breeding season: all year, but eggs often laid toward end of rainy season; number of eggs: 4; incubation period: about 28 days; fledging period: not known; breeding interval: at least 2 clutches per season

LIFE SPAN
Not known

HABITAT
Freshwater marshes and ponds with emergent and floating vegetation such as water lilies and water hyacinths

DISTRIBUTION
Northern Mexico south to Panama; Cuba, Hispaniola and other Caribbean islands

STATUS
Common

bird alone. In addition, observations on the bronze-winged jacana, *Metopidius indicus*, of India, revealed that both adults collected nest material and the female took an active part in courtship, but then left incubation of the eggs and care of the young to the male. The male left the nest only to feed while the female stood guard, driving away moorhens and herons that came too near. Later another male arrived at the pool and, although attacked by the first male, it was able to set up a territory at the far end of the pool. The female would peck the two males when they fought. She mated with the second male and laid a second clutch, which he raised.

Parents distract predators
The chicks hatch out after 28 days or so and are able to run about immediately. When danger threatens, they lie motionless or dive under water while the parents lure the predator away with a distraction display, attracting attention to themselves by flapping their wings and calling.

Some jacanas incubate their eggs by holding them under their wings. This seems strange until it is realized that this will keep the eggs clear of the water when the jacana sits on its flimsy floating nest. The eggs are scooped up with the wings and carried two on each side. After the eggs have hatched, the chicks will also take refuge under their parents' wings.

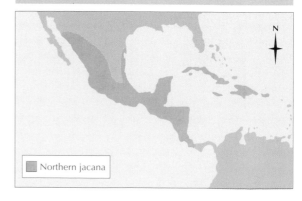

Northern jacana

JACKAL

THERE ARE THREE SPECIES of jackals, all of which are in the dog genus, *Canis*. The three can easily be distinguished by their coloring. The golden or Asiatic jackal, *C. aureus*, is a tawny yellow with black and brown hairs. It has a black-tipped, reddish brown tail and a lighter, near-white underbelly. The black-backed jackal, *C. mesomelas*, is brown gray in color, with a dark saddle of fur along its back and tail, and is pale underneath. Its tail has a black tip. The side-striped jackal, *C. adustus*, is mottled gray, with a pair of light and dark stripes on each side of its body and a white-tipped tail.

The three species of jackals are similar in size. The head and body is about 2½–2⅔ feet (75–80 cm) long, with a 1-foot (30-cm) tail. Jackals stand about 1⅓ feet (40 cm) at the shoulder and weigh roughly 20 pounds (9 kg).

Jackals once ranged throughout Africa, Europe and southern Asia. Nowadays they are found over only half of Africa. The black-backed and side-striped jackals are found in southern and eastern Africa, while the golden jackal lives in northeast Africa. It is only this last species that is still found in Asia and southeastern Europe. Until relatively recently it ranged as far north and west as Hungary.

No competition between species

Jackals are found on savanna and grassland and in arid and open, wooded country. They sometimes come into towns and cities, scavenging for refuse and carrion. They are usually seen singly or in pairs, and are nocturnal animals, foraging mainly at night. Packs of jackals are sometimes seen but are now much rarer than they used to be. The pairs hold territories, sometimes 2 miles (3.2 km) across, marked with urine by both sexes and defended against other jackals. Fights are rare, however, with territorial disputes being settled by aggressive displays.

In the Serengeti, eastern Africa, black-backed and golden jackals have different habitats. Most of the black-backed jackals live in bush country, while the golden jackals live on the open plains. Hence there is little competition. Where the two

Golden or Asiatic jackals with kill, Kenya. Although jackals often scavenge on animals brought down by larger carnivores, they also hunt for their own prey.

The black-backed jackal lives mainly in bush country in southern and eastern Africa. Species differ not only in appearance, but also in some of their feeding and breeding habits.

species do live side by side, as they sometimes do along the borders of the bush, a jackal will allow another of a different species to wander through its territory unhindered. There is also a separation in their feeding and breeding habits.

Hunters as well as scavengers

It used to be thought that jackals were only scavengers. Jackals are indeed carrion eaters and scavengers, following lions, leopards and hyenas in order to feed on the remains of prey left by these larger carnivores. However, jackals are also hunters in their own right. They hunt many small mammals and birds and sometimes take larger animals such as gazelles. They particularly target young gazelles, which they separate from the mothers before chasing and catching them.

The black-backed jackals generally bite at the throat, whereas golden jackals go for the belly. The success of the attack depends very much on whether the jackal is hunting by itself or with its mate. In a study made in the Serengeti, single jackals were successful in only 16 percent of their attacks, while jackals working in pairs were successful in 67 percent of their attacks. Jackals hunting in packs might bring down even larger prey such as antelopes. The attacks are sometimes frustrated by female gazelles coming between the jackals and their prey, butting at the jackals and driving them away. Often food that a jackal cannot eat immediately is carried away and buried to be eaten later.

JACKALS

CLASS	**Mammalia**
ORDER	**Carnivora**
FAMILY	**Canidae**

GENUS AND SPECIES **Side-striped jackal,** *Canis adustus;* **golden jackal,** *C. aureus;* **black-backed jackal,** *C. mesomelas*

WEIGHT
14⅓–33 lb. (6.5–15 kg)

LENGTH
Head and body: 2½–2⅔ ft. (75–80 cm); shoulder height: 1–1⅔ ft. (30–50 cm)

DISTINCTIVE FEATURES
***C. adustus*: mottled gray coat; thin stripe of white and black hairs on flanks. *C. aureus*: tawny coat; darker gray back; near-white underbelly. *C. mesomelas*: brown-gray coat; dark saddle of fur along back and tail.**

DIET
Plant material, insects, rodents, reptiles, birds and carrion. *C. aureus* and *C. mesomelas*: occasionally also larger vertebrates such as gazelles and hares.

BREEDING
Age at first breeding: usually 6–8 months (*C. adustus*), 10–11 months (*C. aureus* and *C. mesomelas*); gestation period: 60–70 days; number of young: usually 4; breeding interval: 1 year

LIFE SPAN
Up to 16 years in captivity

HABITAT
Savanna, grassland and arid, open woodland; sometimes urban areas

DISTRIBUTION
***C. mesomelas* and *C. adustus*: southern and eastern Africa. *C. aureus*: northeast Africa, southeast Europe and southern Asia.**

STATUS
Common

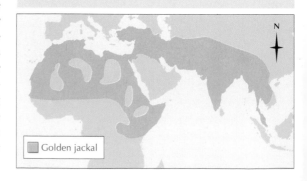

☐ Golden jackal

When feeding in open country, black-backed and golden jackals feed on the same food. This includes young gazelles, insects, carrion, rats, hares, ground-nesting birds and even fruit. In South Africa jackals have become a pest of pineapple plantations, and in India they eat sugarcane and ripe coffee beans. In the bush country, however, the two jackals have dissimilar diets. The black-backed jackals feed more on gazelles and the golden jackal more on insects such as termites and dung beetles. The side-striped jackal tends to be more of a scavenger than the other jackals.

Young born in dens

Jackals give birth to an average of four young, although the range may be between one and nine, after a gestation period of 60–70 days. The pups are born in a burrow that the female digs herself or takes over from another animal and enlarges. Black-backed and side-striped jackals often enlarge holes in termites' nests or old aardvark holes, while golden jackals are more likely to convert the holes of warthogs and other animals to their own needs, often making several entrances. If disturbed, jackals will carry their young to a new burrow.

As is usual in the dog family, there is a strong bond between the parents, and both male and female bring food back for the pups. Whenever one returns there is a ritual: the pups run out with their tails wagging and their ears pressed back. They run alongside their parent, keeping their noses just by the corner of its mouth. The parent that has been guarding the cubs also behaves in this way. The returning parent regurgitates lumps of meat, which are snapped up by the pups or picked up by the second parent and redistributed.

When the pups are 8 months old, they begin to forage for themselves, catching insects, but they may be dependent on their parents for another 2 months or so after this. Golden jackals and black-backed jackals reach sexual maturity when around 10 or 11 months old, side-striped jackals somewhat earlier, at 6–8 months.

Undeserved reputation

"Jackal" and "hyena" are both used as terms of contempt, being synonyms for parasites or sycophants. These animals also have a much exaggerated reputation for cowardice. Such unpleasant associations were derived from the original idea that hyenas and jackals were only scavengers, hanging around the large carnivores for a free meal. Jackals also scavenge around human settlements and have even been credited with robbing graves. Studies that have found jackals to be hunters in their own right have not completely cleared their name.

The jackal's poor reputation has not been helped by their howling in the evening and at night. Their "singing" is thought to be even more disturbing to human ears than that of hyenas.

A black-backed jackal, Kalahari, southern Africa. This jackal may take sheep on pastoral land, where it is treated as vermin.

Index

Page numbers in *italics* refer to picture captions.
Index entries in **bold** refer to guidepost or biome and habitat articles.

Page numbers in *italics* refer to picture captions. Index entries in **bold** refer to guidepost or biome and habitat articles.